KV-638-462

Preface

This volume presents the proceedings of Objective Quality, the Second Symposium on Software Quality Techniques and Acquisition Criteria, held in Florence, Italy, May 29-31, 1995. The first symposium of this series was held in Milan, Italy, in 1994.

As in the previous symposium, the main effort has been devoted to collecting a number of papers covering the various aspects of software quality with special reference to the assessment and improvement of process and product quality. To this end, the knowledge coming from the research institutes and that of the industrial world have been integrated. In fact, the results of the most important European projects in the field of software quality are presented in this book. This will hopefully help to train the symposium partecipants to reach the much yearned-for objective of quality. On this view, the symposium has also been organized following the criteria of quality, in order to provide high quality presentations and proceedings. In this respect, I would like to express my sincere thanks to the authors of the papers and to the committee members.

One keynote speaker was invited to present the impact of quality technologies and to show the way to future evolution. To this end the choice of Prof. G. Bucci was applauded since most of the work in software quality in the Florentine area is the result of his early initiatives.

During the symposium three additional events have also been organized. The first two are devoted to the presentation of the IBM and Olivetti activities, respectively, in the field of quality. The latter concerns to the presentation of the UNINFO (Italian National Body of ISO) standards collection. Moreover, vendors' exhibitions, and demonstrations of new products and literature have also been offered. Thus, I trust that researchers and practitioners will find several opportunities for exchanging their ideas and for following the path to reach the objective of quality.

Finally, I would like to thank CESVIT (High Tech Agency), Area della Ricerca C.N.R. (Consiglio Nazionale delle Ricerche), QualityLab consortium, and the Department of Systems and Informatics, University of Florence, for their organization; the sponsors for their support, and the members of Consulta Umbria for their high quality work. A particular acknowledgment to Carla Pardini and Paola Lumachi for their help. Many thanks to the participants to the symposium who really make it exciting. Lastly, I would like to thank Paola and Davide for their patience and understanding.

Florence, March 1995

Paolo Nesi
Objective Quality 1995 Chairman

Organization

Objective Quality 1995 is organized by: CESVIT High-Tech Agency, Florence; Area della Ricerca CNR (Consiglio Nazionale delle Ricerche), Florence; QualityLab consortium, Turin; and the Department of Systems and Informatics, University of Florence, Italy.

Supporting Institutions

Banca Toscana
IBM Semea
Il Sole 24 Ore
Olivetti
SUN Microsystems Italy

Promoting Institutions

De Qualitate
Distretto Tecnologico del Canavese
Hi-Tech Network (Tuscany Network Hi-Tech)
Pmi Market & Marketing
Qualità (AICQ, Italian Association for Quality)
Qseal (Quality Seal, certification consortium)
TABOO (Italian association on advanced technologies based on object-oriented)
Toolnews
UNINFO (Italian National Body of ISO)
ZeroUno (Mondadori Informatica)

Executive Committee

Chairman: P. Nesi (University of Florence, Italy)

Committee: M. Campanai (CESVIT/CQ_ware, High Tech Agency)
A. Cicu (QualityLab consortium)
A. Serra (ASIC)
A. Tronconi (Area della Ricerca CNR)

QA66.5 LEC/926

Lecture Notes in Com

Edited by G. Goos, J. Hartman

Advisory Board: W. Brauer D. Gries J. Stoer

WITHDRAWN
FROM STOCK
QMUL LIBRARY

DATE DUE FOR RETURN

NEW ACCESSION
CANCELLED

10 JAN 1997
- 2 DEC 1999
16 JAN 2006

Springer
Berlin
Heidelberg
New York
Barcelona
Budapest
Hong Kong
London
Milan
Paris
Tokyo

Paolo Nesi (Ed.)

Objective Software Quality

Objective Quality: Second Symposium on
Software Quality Techniques and Acquisition Criteria
Florence, Italy, May 29-31, 1995
Proceedings

Series Editors

Gerhard Goos
Universität Karlsruhe
Vincenz-Priessnitz-Straße 3, D-76128 Karlsruhe, Germany

Juris Hartmanis
Department of Computer Science, Cornell University
4130 Upson Hall, Ithaca, NY 14853, USA

Jan van Leeuwen
Department of Computer Science, Utrecht University
Padualaan 14, 3584 CH Utrecht, The Netherlands

Volume Editor

Paolo Nesi
Department of Systems and Informatics, University of Florence
Via S. Marta 3, I-50139 Firenze, Italy

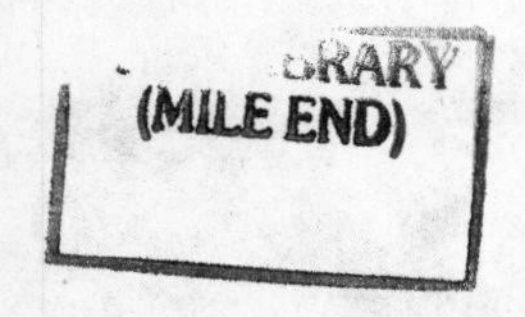

CR Subject Classification (1991): D.2, K.6.3-4

ISBN 3-540-59449-3 Springer-Verlag Berlin Heidelberg New York

CIP data applied for

This work is subject to copyright. All rights are reserved, whether the whole or part of the material is concerned, specifically the rights of translation, reprinting, re-use of illustrations, recitation, broadcasting, reproduction on microfilms or in any other way, and storage in data banks. Duplication of this publication or parts thereof is permitted only under the provisions of the German Copyright Law of September 9, 1965, in its current version, and permission for use must always be obtained from Springer-Verlag. Violations are liable for prosecution under the German Copyright Law.

© Springer-Verlag Berlin Heidelberg 1995
Printed in Germany

Typesetting: Camera-ready by author
SPIN: 10486208 06/3142-543210 - Printed on acid-free paper

Contents

Quality and Users

Product Quality

Testing

Objective Quality, 1995

Paolo Nesi

Department of Systems and Informatics, University of Florence
Via S. Marta 3, 50139 Florence, Italy
Tel: +39-55-4796523, Fax: +39-55-4796363
email: nesi@ingfi1.ing.unifi.it, www: http://www-dsi.ing.unifi.it/~nesi

General Introduction

Software quality is becoming the key issue in the field of information technology for maintaining and/or proposing products and, thus, industries on the market. On the other hand, since quality is for many regards a subjective issue, its meaning and achievements remain quite unclear. The most widespread definition of quality specifies that its degree directly depends on customers' satisfaction, while another definition is based on the statement "fitness for the purpose". In effect, both these definitions are reductive since quality is something which is more and more complex, and varies from an industry to another. In addition, quality needs to be modeled at least on the basis of customers' needs, context of the market, industrial goals, system requirements, etc. Therefore, quality profile, specifying the importance of each quality parameter, must be defined by also considering the effect of these factors. As regards quality achievement, most of the major industries are quite convinced that by increasing products quality a sensible reduction in costs for systems software design, maintainability, etc., is obtained; on the contrary a great part of small and medium-size industries are not very sensitive to the quality problems yet. This is due to the fact that (i) the mechanisms by which product quality can be achieved and how quality criteria can be introduced in the company, (ii) to what extent product quality must be obtained in order to meet the industrial goals, and (iii) what are the measurements should be performed for a correct evaluation of software quality, etc., are still unclear. Moreover, these factors are less clear to the industries which are not used to perform long-term planning and auto-assessment. Therefore, the uncertainties related to the lack of answers to the above mentioned questions mainly depend on the status of the industry in terms of basic know-how, and flexibility of mind-set management.

These facts justify the organization of a Symposium named Objective Quality. This name has a double valence, objective as a yearned quality and a non subjective measure of quality. Therefore, during the selection of papers much attention has been devoted to provide guidelines and possible solutions to the above mentioned problems. Moreover, the following introduction has been added in order to explain the relationships among the papers presented and between these and the rest of the technical literature, and not to provide an omnicomprehensive survey on quality issues.

Overview Description of Papers

The contributions that have been presented at the symposium can be categorized into roughly five groups plus a keynote.

Bucci, in his keynote paper, explains the lessons learned on the basis of his experience in introducing and diffusing the concepts of software engineering and software quality in several industries. This paper highlights the criticism and points out both the path for further works for researchers and the tendencies for industries. To this end, the criticism related to the introduction and assessment of software quality process, software development, testing, etc. in the context of the actually applied technologies is reported. A special attention is devoted to consider the points of view and the needs of managers, with respect to the results which could be really achieved, and those of the purchasers with respect to quality certifications.

Process Assessment and Improvement

For the introduction of quality criteria in a company, quality models and techniques for modifying the process of development and for its control have to be adopted. The process of transformation is obviously neither immediate nor easy to be applied. Moreover, it should be performed without producing discontinuities in the company production since these usually lead to discontinuities in product quality, decreasing in efficiency and increasing of costs, etc. For these reasons, the quality improvement in the development process of the company must be always supported by techniques for the assessment of the process quality according to companies' quality requirements and business targets. For controlling the process of quality improvement an assessment method is obviously needed. This can be based on objective (instead of subjective) measures and/or on questionnaires. In the context of software quality, in order to guarantee a uniform level of quality among developers, and thus to protect the end users, several standards have been defined by international and national organizations [1]. These contain guidelines for describing the use of quality characteristics for the evaluation of software quality processes (e.g., ISO 9000, UNI EN 29000, series, DoD 2167). For these reasons, most of the approaches for process assessment and improvement explicitly declare their compliance to standards.

Kuntzmann-Combelles's paper describes real experiences of software improvement based on the **ami** (Assess, Analyse, Metricate, Improve) approach for metrication. The **ami** approach has been partially supported by the European Community, and is based on CMM of SEI (Capability Maturity Model of Software Engineering Institute, based on DoD2167) assessment method and on the GQM (Goal/Question/Metric) paradigm [2], [3], [4]. The CMM improvement process is comprised of 5 sequential levels of organization capability of software development [5]. The levels are: initial, repeatable, defined, managed and optimizing. The goal is obviously to reach the optimizing level if possible with an acceptable effort. The **ami** approach is capable of tuning the CMM approach according to the business needs of the company in order to better focussing the efficiency of the improvement approach. The **ami** approach is comprised of 12 steps which ensure benefits for project planning and management, cost-effectiveness and matching of business goals. The experiences proposed in the Kuntzmann-Combelles's paper follow a GQM paradigm for mixing business goals defined by the top level management with the results of the CMM SEI assessment in order to define action lists and associated measurements to follow the progress of the actions. Experiments reported in this paper have been exploited in industrial European sites. From these experiences, some useful lessons have been learned and thus reported in this paper.

Kuvaja's paper presents the BOOTSTRAP approach for software process assessment and improvement. The definition of this approach has been partially supported by the ESPRIT program of the European Community. It offers a methodology which can be suitably used with different types of the software processes and products, and with different software producing organization structures and sizes. This approach is supported by a cumulative data base for the software industry as well as for each individual sub-sectors. BOOTSTRAP is based on the ISO 9000 series quality standards and on project management standards, ESA-PSS-05 (European Space Agency). It extends the CMM of SEI in many directions, providing (i) a higher integration between assessment and improvements, (ii) estimation of maturity level for each process attribute, (iii) managing 30 additional quality attributes. Moreover, it supports several standards, DoD 2167 (prescriptive) and ESA (descriptive); and takes into account the methodology for software development chosen. This paper presents an evaluation about the performance of BOOTSTRAP approach in Europe in the last years and its further developments.

Coletta's paper reports an overview about the SPICE (Software Process Improvement and Capability dEtermination). It is due to an international effort for developing a Standard for Software Process Assessment supported by the International Committee on Software Engineering standard, ISO/IEC JTC1/SC7/WG10. Several software organizations belonging to 16 countries, including the major suppliers of Software Process Assessment methods such as SEI (promoter of CMM) and the BOOTSTRAP consortium, are supporting the project. The process assessment method under development will provide requirements and guidelines for conducting the process of assessment by using a model of processes and the rules of good software engineering. It also provides guidance on Process Improvement, Capability Determination, Assessors Training and Qualification, and use of Assessment Instruments. This project is now undergoing world-wide trials.

Campanai's paper presents some lessons learned in introducing quality control in the process of software development for banking applications. The approach adopted for process improvement and assessment has been customized on the basis of several approaches in order to get the best results in a shorten time and for integrating techniques which are peculiar to management. The method proposed for software process improvement has been directly defined on the basis of the company needs and, thus, several adjustments with respect to the traditional and well-known approaches for process assessment and improvement have been set. The lessons learned, and explained in this paper, are quite general to be applied in other fields, especially when model tuning is mandatory; customized approaches are often better accepted when the quality process control is introduced *ex-novo* reducing in this way the acquisition effort, or when other approaches have not been completely satisfactory.

Software Modeling and Quality

In order to improve process quality, the development process itself must be changed in many instances. Well defined and constrained methodologies, and strongly formal approaches (operational or denotational, descriptive) for software development can be profitably used in order to improve repeatability and human-independentness of system production quality [6], [7]. A higher confidence can be reached if such methods are supported by mathematical foundations providing mechanisms for verifying specification consistency and completeness, and for validating system behavior with respect to initial requirements. To this end, several CASE tools and formal languages have been proposed [8]. Such an approach to software specification is practically mandatory when the systems under development have to satisfy real time constraints. In fact, in such conditions the formal verification and validation often allow to identify critical conditions and, thus, to improve system quality.

Bruno and Agarwal demonstrate the efficiency of the validation process for obtaining high quality specifications. Their paper is based on Protob and Quid model languages which are well-known and documented in the literature. Protob is an object-based language and covers functional and behavioral issues, while Quid covers the informational aspects. Both these models are operational and their specifications are based on formal models which can be simulated and animated highlighting the dynamic aspects of the systems modeled. The operational semantics is guaranteed by an extended Petri Net model. The system simulation allows the validation of the specified system with respect to its initial requirements.

Lofrumento and Pileri discuss in their paper the design quality achieved by using the approach named Object Design Method (ODM). ODM is an object-oriented method supported by a corresponding formal model which extends both the OMT methodology proposed by Rumbaugh *et al.* [9] and the Object Specification Method proposed in the literature by Lofrumento. Its extensions have been mainly focussed on better specifying communications and concurrency among objects as they usually occur in real-time and distributed systems. This makes this approach particularly suitable for designing telecommunication systems such as that reported in this paper. Lofrumento and Pileri also discuss their model and the related methodology in view of the ISO 9126 quality concept.

Reuse

If the development process method provides an efficient support for software reuse a reduction of costs of system development is obtained. At the same time, if the components reused have a high quality, an improvement of the overall system quality could be theoretically achieved. This usually happens since the components reused have been already verified, validated and tested for several other projects in which they have been used; thus, their quality is easily high. On the other hand, the reuse process to be profitably used needs to be supported by: *formal mechanisms* in order to automatize the retrieval of reusable components; *metrics* to (i) predict the costs of adaption of the components under reuse, (ii) predict the final software quality, and (iii) evaluate the intrinsic reusability of components, etc.; and *guidelines* for controlling the process of development *with* and *for* reusability.

The following two papers explain the main experiences learned from the ESPRIT project REBOOT (REuse By Object-Oriented Techniques). The first is mainly focussed on what can be reused and which characteristics the components must have to be reusable, while the second describes the impact of software reuse on quality.

Stålhane's paper reports the foundations on which the REBOOT reusability assessment model has been built. This model has been defined by identifying what is considered important for reuse by software engineers, and allows to assess the components reusability. In order to define such a model, a questionnaire was defined and filled up from expertise of Norway, Sweden, Germany, France and Spain. The questions were grouped according to three main aspects which are: problem, components and production features. Then, the questions were re-classified on the basis of Factor - Criteria - Metrics model (FCM) of IEEE standard in order to have a support for interpreting the answers as measures of understandability, confidence, and portability, and their corresponding sub-characteristics. The results of this evaluation have highlighted what makes software components reusable and how reusability could be evaluated. The model for reusability assessment has been implemented as the qualification tool in the REBOOT toolkit prototype.

Karlsson and Morel present in their paper an analysis about the impact of reuse on quality. More precisely, it is shown that the concept of quality, according to the ISO 9126 definition, includes different aspects which are influenced by reuse: functionality, reliability, usability, efficiency, maintainability, and portability. This paper, firstly discusses in detail how the reuse can help to improve these factors, then a discussion about the development *for* reuse approach vs. development *with* reuse is given, addressing the problems of widening, narrowing, isolation, configurability, etc. On these bases, guidelines for defining and improving a design for reusability process are proposed, in order to effectively introduce reuse in the software development process for managing organisational, managerial and technical aspects of reuse as a coordinated evolutionary process. Finally, a study on how to incorporate reuse in the software acquisition process is reported.

Sommerville, Masera and Demaria present the results of the European ESSI project APPRAISAL. This project has been focussed on the definition of practical reuse guidelines which can be applied without introducing an unacceptable increase in development costs, thus following a design *for* reuse approach. This study has been performed for covering the lack of efficient guidelines for producing reusable Ada components, since most of those presented in the literature can be applied with difficulty in industrial environments. The proposed guidelines must be adopted as a part of the development process in order to increase components flexibility and compliance with the requirements of ideal reusable components. In order to make available the reusable components, these have been organized in an archive which can be browsed by a local World-Wide Web (WWW) engine, since this approach quite satisfies the requirements of a reuse dissemination tool. Obviously, a WWW-based system cannot exploit formal approaches for reuse and complex organizations of reusable components, but it consists in a simple and efficient multimedia tool for navigating in a distributed hypertextual archive in which code, documents, pictures describing reusable components are employed. Moreover, reusable components can be made easily available since the database can be distributed and the relationships of hypertext locally defined.

Quality and Users

In this section, two different problems which have in common the fact that in both cases much attention is devoted to the user's point of view. The first analyzes the user's view with respect to the problem of process improvement, while the other evaluates system usability. This feature is becoming one of the most important aspects of product quality. This fact is also due to the diffuse adoption of sophisticated graphic multimedia user interfaces.

Cicu's paper is focussed on discussing how the assessment and improvement models are considered by the user/purchaser. In this respect, the models SEI, SPICE and the standard ISO 9000, AIPA (Italian directive for public administrations), are analyzed for comparing their efficiency in supporting the interests of the user/purchaser; in particular: the costs/benefits in performing the improved process, how software functional and quality requirements are related to the purchaser's needs, how the users/purchaser can be supported for contractual and collaborative relationships with the supplier. Additional notes on IEEE standard, ami are also given.

Binucci's paper reports an overview of the European ESPRIT project MUSiC. This project is based on the observation that the concept of usability involves not only software, machines and documentation, but also people who obviously interact with them. The MUSiC method allows the estimation of usability considering users' satisfaction, performance, understandability, as well as a specific theory. The four measures can be used in different phases of software life-cycle, and are supported by a corresponding tool. In the same paper, an analysis of cost reductions due to the application of the model proposed is reported.

Product Measurements

According to ISO 9126 the product quality can be estimated on the basis of its functionality, reliability, usability, efficiency, maintainability, and portability [2], [10]. Moreover, other additional more specific product metrics can be defined in order to evaluate product reusability, understandability, etc. On the other hand, the importance of each of the above mentioned factors with respect to the whole product quality in a certain context can vary to a large extent. For this reason, the so-called product quality profile is usually defined on the basis of company experiences or by applying the GQM paradigm to a larger set of experts. Despite this fact, the estimation of product quality could be all the same a very difficult task. Code measures, defined as indirect index for the above mentioned factors, may depend on these factors following undefined relationships which in turn may depend on the programming language and on the development process itself. Moreover, the definition of orthogonal metrics which are capable of estimating indexes corresponding to the above features without being influenced by other features is up to now unrealistic. These reasons make the world of product metrics very huge and complex to be depicted without the support of formal techniques and the adoption of rigorous methods for their definition and application [11].

Bakker and Hirdes report in their paper results obtained by the experience due to the analysis of several millions of lines of code belonging to the field of communications, banking, insurance, railway, government, etc. This code was written in high-level languages (such as C, C++, COBOL, Pascal, etc.) and has been analyzed for different reasons – i.e., redesign, restructuring, quality evaluation, testing, certification, costs evaluation, etc. The methodology applied is based on the ISO 9126 and is implemented as the COSMOS Workbench. From this analysis they have learned that (i) the main problems of systems legacy is due to their design and not to code structure; and (ii) redesign and restructuring are the less costly and safer approaches for improving system quality than system rebuilding *ex-novo*.

Spinelli, Pina, Salvaneschi, Crivelli, and Meda present their experience in measuring quality of large systems. The approach adopted by these authors extends the concept of quality profile, and consists in building a quality matrix, which relates a quality profile to each single functionality of the product, to apply a pruning technique for cutting the measurements which should not be taken into consideration during the whole product quality assessment. The approach proposed allows to evaluate large systems with a lower effort with respect to traditional techniques. Since these usually have a high complexity of measurements, with the approach proposed in this paper a strong reduction of measuring costs is obtained.

Testing

In most of the classical software development life-cycles, the testing phase is usually located at the end of the process after the formal validation. New policies and approaches for testing attempt to distribute tests along all phases of software development process with different degrees of detail for better detecting different sources of faults. The first two papers in this section are focussed on proposing testing approaches integrated and distributed in time. Moreover, software testing, instead of for fault identification, can be also used to produce useful information about the system reliability (i.e., dependability) as depicted in the last paper. This is very useful since the reliability is in many contexts one of the most important factors of product quality.

Bazzana, Delmiglio, Lora, Balestrini, and Finetti describe in their paper the experience in introducing testing at Siemens Telecommunications. The tests have been introduced by their integration in different phase of the development process. Quantitative results from a systematic integration test on parts of a GSM (Global System for Mobile communications – cellular phones) Phase 2 system are obtained. The impact of test integration has been evaluated with respect to project management and the product quality, and for the derivation of baselines (for further use), and for maintaining under control the improvement due to the introduction of test integration. The results show a positive impact of the approach proposed with respect to product quality, highlighting the importance of anticipating tests on the development cycle.

Bruno, Varani, Vico, and Offerman report in their paper the experiences in using a model-based approach for testing. This approach consists in modeling the external environment of the equipment under test; such a model must be capable of providing inputs for the systems and for verifying in time the corresponding answers. To be cost effective the modeling must obviously be done with an efficient and flexible tool. In this case, Protob operational model has been used; this allows to build descriptions which are directly executable. In this paper, three experiences about the technique described are reported. The results clearly show that the costs related to model development are widely covered through the savings in test execution time and product quality.

Bertolino's paper shows how the conventional reliability-growth model can be used for predicting the software reliability by estimating the failures observed during testing. Reliability is an important parameter of product quality; moreover, a measure of system reliability can used for mathematically evaluating the probability of failure and, thus, product safety and quality. Critical conditions in which software does not fail during testing are at the current state of the art unmanageable. These cases correspond to the safety-critical applications in which the ultra-reliability is mandatory. As an alternative approach, Bertolino, shows that a measure of software testability can be used for predicting system dependability.

References

1. S. L. Pfleeger, N. Fenton, and S. Page, "Evaluating software engineering standards," *Computer*, pp. 71–79, Sept. 1994.
2. V. Basili and D. M. Weiss, "A methodology for collecting valid software engineering data," *IEEE Transactions on Software Engineering*, vol. 10, no. 6, pp. 728–738, 1984.
3. V. R. Basili and H. D. Rombach, "Tailoring the software process to project goals and environments," in *Proc. of the 9th International Conference on Software Engineering,*, (Monterrey, CA, USA), pp. 471–485, IEEE Press, March 30-April 2 1987.
4. V. R. Basili and H. D. Rombach, "The tame project: towards improvements oriented software environments," *IEEE Transactions on Software Engineering*, vol. 14, pp. 758–773, June 1988.
5. M. C. Paulk, B. Curtis, M. B. Chrissis, and C. V. Weber, "Capability maturity model for software," tech. rep., Software Engineering Institute Carnegie and Mellor University, Ver.1.1, CMU/SE-93-TR024, USA, Feb. 1993.
6. B. W. Boehm, "Software requirements economics," *IEEE Transactions on Software Engineering*, vol. 10, pp. 4–21, Jan. 1984.

7. B. W. Boehm, "A spiral model of software development and enhancement," *IEEE Software*, vol. 21, no. 5, pp. 61–72, 1988.
8. G. Bucci, M. Campanai, and P. Nesi, "Tools for specifying real-time systems," *Journal of Real-Time Systems*, pp. 117–172, March 1995.
9. J. Rumbaugh, M. Blaha, W. Premerlani, F. Eddy, and W. Lorensen, *Object-Oriented Modeling and Design.* New Jersey: Prentice Hall International, Englewood Cliffs, 1991.
10. N. E. Fenton, *Software Metrics: a Rigorous Approach.* London: Chapman and Hall, 1991.
11. N. Fenton94, "Software measurement: A necessary scientific basis," *IEEE Transactions on Software Engineering*, vol. 20, pp. 199–206, March 1994.

The Impact of Software Quality

Giacomo Bucci

Dipartimento di Sistemi e Informatica, Università di Firenze
via S. Marta 3 – 50139 Firenze, Italy

Abstract. This note reports a number of considerations on Software Quality, based on the experience made in recent years in the context of Technology Transfer initiatives. It is argued that, with the current state of the art in Software Technology, efforts should be primarily devoted to add more formalism and rigour in the whole development process by also considering and integrating all its aspects. Moreover, standardization of metrics and harmonisation of certification are also mandatory. These, in turn, should promote the establishment of organizations whose members agree at least on a minimum set of metrics and certification criteria.

1 Introduction

Software is pervading, in a direct or indirect manner, almost any aspects of our life. Software is central to many industrial products. It may contribute to the improvement of existing products in terms of added capabilities and reduced costs; it also facilitates the development of new, less expensive and more efficient products. Software is instrumental for the greatest part of production processes, being the core technology for controlling, monitoring and reporting activities. Software is essential in banking, administration and any sort of business activity. It is also essential in services like telecommunications, transportations and the like.

As society becomes more dependent on software, there is a growing demand for improving its quality. This explains why software quality has become a great concern for both software producers and users. The search for improving software quality has, in a general sense, the same motivations and the same expectations as for any other product. It is already clear that qualification of software quality is becoming fundamental for staying in a competitive market and for the product internationalization. This has two aspects:

- the compliance of the production process with some standardized quality requirements (e.g., ISO 9000);
- the certification of the quality of the software product.

As it has been many times stated, the production of software has been considered for a long time as work of art. The search for higher quality is one of the compelling reasons to move from a craftsman activity towards a well-organized industrial process. Of course, other industrial or service sectors have been facing the problem of quality management since the long time and a consistent body of knowledge and practice was available which has been applied to the software production process. Several models for this process have been proposed, as well as techniques for its assessment (some of which are the results of projects funded by the European Community). International organizations like ISO have produced a set of requirements and guidelines for dealing with quality in general, and with software in particular. The emphasis is on *the quality system.* This includes process organizational structure, responsibilities, procedures, resources and any activity that influences the quality of the software product. It is assumed that process compliance to the requirements of a quality model is a guarantee for the quality of the product (i.e., software).

The above issue is not always true, and thus objective metrics for evaluating the product are needed. In this scenario, it is likely that, for the years to come, metrics will be still largely subjective, unless rigorous (and widely applicable) methods for quantifying software quality are found. This imposes that the various actors in software quality (producers, users and certifiers) agree on a minimum set of metrics, so as to satisfy the needs of all of them.

This note pays little attention to the *process,* focussing on the evaluation of *product* quality. It draws from the experience gained by the author while collaborating in CESVIT (High Tech Agency, Italy) activities; the University of Florence has a participation in CESVIT. In particular, with CQ_ware (Center for Software Quality) established by CESVIT 1990. The first objective of CQ_ware was to help the local community in raising its capabilities in software engineering. The experience is based on re-engineering systems, introducing quality improvement and assessment techniques, testing, product evaluation for more than 300 electronic Small and Medium Enterprises (SMEs) that can be considered High Tech according to the European Community parameters, plus banking, software houses, etc. A 46 per cent of the mentioned SME companies generate significant amount of sales in export. Most of them produce electronic control systems or electronic equipment in which software has already become the most critical component.

2 Software Engineering and Software Quality

In a broad sense, the obvious approach to improve quality is to start with good engineering practices. In fact, quality is something which cannot be added afterwards. It is the result of the entire production process. In particular, there is a general agreement in considering the early project stages, including specification of requirements, as the most important phase for the final product quality.

Software engineering has essentially evolved as an empirical discipline. As a matter of fact, the most popular methodologies are rather informal. CASE tools supporting those methodologies are employed in many production environments.

The hope of every practitioner is that the use of those methodologies will result in better products. This is true to a certain extent, but it does not guarantee that the outcome is always as good as required. For instance, nobody can guarantee that the implementation in a traditional language (e.g., C, Fortran, Pascal, Cobol, etc.) after a *perfect* analysis and design phase (done, for instance, through Data Flow Analysis or by other means) will produce a high-quality system. This is much more stressed when specific languages (such as VDM, Z, etc.) are used in the phase of requirements analysis when a methodology is selected to draw the analysis and design phases and finally an unrelated language is adopted for coding. In these conditions, much work is performed in integrating tools, through testing procedures and methodological manuals. But unfortunately, to say it in other words, there is no guarantee that having chosen all the right ingredients the supper will taste good.

Another key aspect is that, though many firms have developed their own quality system, the special nature of software, and the lack of rigorous quantification techniques are such that software quality remains a subjective matter.

This would not be the case if software technology had reached a stage in which the entire software production process would be carried out in a formal, rigorous and integrated manner without boundaries among the several process phases. Much research is done with this objective in mind. Formal languages have been proposed which can be applied at different specifications levels with also the capability of covering with the same model a wider part of the process. In principle, a formal language allows predictability of system behavior, as well as requirement validation by means of property proofs at any development phase. Rigorous techniques would also imply rigorous metrics. These must consider the quality with a wider view with respect to the traditional quality profile.

Unfortunately, the above described is not the state of art. Besides, formal techniques have proved to be difficult to be used, given the somewhat complex mathematical apparatus that they need. As a result they are hardly used in real software production environments. Moreover, not much work has been done for providing metrics for that techniques.

3 Software Quality

As the software market grows, users get conscious that software procurement should be based on objective criteria rather than on the push of vendors. Mature users have begun to impose that what they buy must be certified. This means certification of compliance of the production process with a given quality model (e.g., ISO 9000) and certification of software quality.

As already noticed, compliance of the production process is important and, in the lack of a certification of product quality, it is an indirect guarantee for the end user. However, end users would be much more satisfied by a direct, quantitative measure of the product quality especially if this is demonstrated by considering their point of view.

ISO 9126 identifies a number of Quality Characteristics and provide guidelines for their use. They are a definite step for ruling relationships between the producer and the user. The method for evaluation is based on: (i) the definition of an expected quality profile (based on answers to questionnaires); and (ii) the comparison of that profile with the measured quality profile, obtained through functional tests and the use of metrics.

Please note that the identification of the expected quality profile is of great importance, since higher profiles imply higher costs, and wrong profiles imply higher values for the cost/benefit ratio. The techniques for defining quality profiles should be improved by considering the user's needs and point of view, context of the market, cost/benefit ratio, company experiences, and not only the ISO 9126 model.

4 Quality Evaluation

On the basis of the author's experiences, several open issues have been identified, including:

- harmonization of certification/evaluation;
- homogeneous metrics for different stages of software life-cycle;
- quality metrics.

4.1 Harmonization of certification/evaluation

Harmonization of a certification scheme is an open issue both for customers and users (to be sure that they are paying for the right price) and for producers (to prove their capabilities). The definition of a common agreement on unique meanings for quality in general and for metrics in particular, as well as the definition of their acceptable ranges is a real obstacle to certification. In these cases, the solution could be to define a minimum set of metrics covering: (i) the technical and functional aspects of the software product, and (ii) the characteristics of software process development.

As regards the technical aspects of the software product, customers/users and software developers must agree on:

- a taxonomy of software quality (e.g., ISO 9126 or a wider customarily defined one);
- a well-defined set of characteristics that can be measured based on that taxonomy;
- a measurement unit for each characteristic;
- a well-defined set of procedures for collecting data.

As to the functional aspects of the software product, customers/users and software developers must agree, at least, on: (i) quality of the requirements (this is still a real issue for research), and on (ii) classification of functional requirements for classes of applications. In this respect, public institutions (government, military, etc.) should promote the definition of such requirements.

Regarding process characteristic, there exists the need of going beyond the certification of the quality system, so as to take into account also the quality of the development process. For instance, it is much more important to demonstrate an efficient software maintenance than to attest that the performing of maintenance is well documented. Of course, the implication is that quality is not simply a seal that you present to your costumers.

4.2 Homogeneous metrics for different stages of software life-cycle

In general, there is the need for general purpose metrics to be applied to the entire life-cycle, with specific characterization for the individual stages of the life-cycle. To define *general purpose metrics* means also that the results obtained by using these metrics must be interpretable with a uniform semantics independently of the context (language, process phase, market, development methodology, etc.) in which they have been estimated.

For instance, if we consider the general purpose metric *software volume*, what is needed is a unique measurement unit, even though different specific metrics are applied to the various stages. These will produce homogeneous results, making them comparable. In addition, this model will permit validation, phase by phase, of the entire life-cycle, on the basis of measurements on software products. Different results and applications of the same metric concept on distinct contexts must be comparable on the basis of their estimated values, ranges and trend.

Furthermore, there is the need for embedding metrics within the software development environment (i.e., the software factory). In other words, future CASE tools should measure product quality at each stage of the development process, by automatically gathering data on the basis of implemented metrics. Please note that such a feature seems to be very important for SMEs, since they usually do not have enough resources for measuring software along the phases of its development. On this view, also the environment for software development (CASE tools, workbenches, etc.) must be modified for considering the metrics evaluation, allowing the definition of quality models and event triggers on their values.

4.3 Quality metrics

A superabundant number of software metrics has been proposed in the literature, most of which have little or no scientific foundations. However, industrial practice shows that empirical metrics are often preferred. These are frequently generated by the experience and, thus, summarize the history of the company with a suitable abstraction. On the other hand, the rigour of the measure theory could improve their efficiency by also avoiding trivial errors.

There is a great difficulty in validating metrics. Quality software metrics have been often based on statistical approaches, which imply stable and modelable conditions. Unfortunately, the technological context is rapidly changing, thus questioning the validity of the results statistically obtained in time. For instance, consider the case of a language/compiler update, going from version V_i to V_{i+1}.. A statistical model, validated for version V_i does not guarantee that, when applied to version V_{i+1} that valid results are obtained.

The definition of validation approaches which are both scientific and flexible enough to be adapted to the changing context is an open issue. Moreover, the validation context is often not very clear and reproducible; thus, the validations proposed have in many cases no sense. The validation is usually made by using reference measurements; these are often too much subjective to be considered an effective reference (e.g., a restricted number of measures, measures obtained in wrong and uncontrolled contexts, a posteriori re-estimations, etc.). The validation should also be extended to define a predictable profile metrics with respect to the software life-cycle. This could be very useful for defining process-oriented quality control environments.

Software evaluation should produce consistent, repeatable results, in the sense that it must have a coherent output independent of the evaluator, methodology, language, etc. A common set of standard metrics must be agreed upon, as well as the criteria for applying them. As a conclusion, software certification makes sense as long as it really gives a real measure of quality.

Quantitative Approach to Software Process Improvement

Annie KUNTZMANN-COMBELLES
OBJECTIF Technologie, France

Abstract. The paper describes real experiences of software process improvement based on the SEI CMM assessment method and the **ami**[1] approach for metrication. The **ami** method is an established technique an organization can use to get started with process improvement. A 12 steps supported approach ensures benefits for project planning and management, cost-effectiveness and match of business goals. The **ami** project has been partly funded by the CEC and led by practitioners in the European Software Industry. The method takes benefit of the Goal-Question-Metric (GQM) paradigm and mixes business goals defined by the top level management with the results of the SEI assessment in order to define action lists and associated measurements to follow the progress of the actions. Experiments have been running at industrial sites and some of the most useful lessons learned during these projects are reported here.

1 introduction

a m i (ami stands for **A**ssess, analyse, **M**etricate, **I**mprove) was developed in Europe as a cooperative project involving nine European centres of excellence and is an instance of applying two processes techniques familiar to Americans: the Goal-Question-Metric paradigm and the SEI"s CMM. The aim of the project is to provide a practical and validated approach to installing and using quantitative approaches to control and improve the software process. The **a m i** approach has undergone extensive industrial trials across Europe first during the two years project step and furthermore in various industrial contexts where software process improvement had been identified as main need and policy for the next 3 years of a company. In every market segment - avionics, railways systems, telecommunications or business applications - **ami**'s usefulness has been recognized in determining the right decisions concerning processes.

The reason for process management strategy is most of the time driven by the business competition and the increase of the software part in the marketable products. Whatever the complexity of the product is, software part becomes one of the main factors for added value and as a consequence, management control and improvement of the software process are mandatory.

---

[s] **ami** was developed by a consortium consisting of GEC-Marconi Software Systems, ALCATEL Austria Forschungzentrum, BULL AG, OBJECTIF Technologie, GEC Alsthom, ITS, RWTÜV, South Bank University and O. Group and supported by the CEC under the ESPRIT initiative.

2 Summary of the ami method

Once process problems have been identified - and software process assessments are one of the best ways to do so - an organisation must carry out some key activities successfully in order to make major improvements in its software process. It must

- create a translation into business goals of CMM issues,
- derive improvement goals,
- develop an action plan to achieve those goals,
- define and use effective metrics to track progress towards those goals during the many months - perhaps years - required for a major improvement step.

Industrial experience has demonstrated the usefulness of two different approaches to process improvement: analysis and benchmarking. the analytical approach relies quantitative evidence to determine where improvements are needed and to judge the real improvement achieved. The benchmarking approach depends on identifying excellent practices in one organisation and introducing them in your context, assuming that what is excellent elsewhere remains excellent for you. the SEI CMM is an example of a benchmarking approach for software process. Our experience is that no one of these methods is efficient if applied independently: coupling the two approaches into an integrated framework provides a better cost/benefit ratio and ensures that business goals are matched. The **ami** approach supports this integrated framework.

The application of goal oriented measurement in an organization requires a structured method. Each organization must construct its own improvement framework. Which organization, after all, would borrow the mission statement of another?

The **a m i** method implements four distinct activities - **A**ssess, **A**nalyse, **M**etricate, **I**mprove :

Assess your project environment (with its objectives and problems) to define primary goals for measurement. Managers who initiate measurement must be involved in this activity.

Analyse the primary goals to derive sub-goals and the relevant metrics. This analysis is formalised as a goal tree with a corresponding set of questions to which these metrics are linked. The participants affected by the metrication goals (metrics promoter, project managers, quality engineers, etc.) will generally carry out this activity.

Metricate by implementing a measurement plan and process the collected primitive data into measurement data. The metrics promoter will write the measurement plan and coordinate its implementation.

Improve, as the participants affected by the goals start to use the measurement data and implement actions. Comparison of the measurement data with the goals and questions in the measurement plan will guide you towards achievement of your immediate project goals. When your measurements show that you have achieved a goal, you have improved enough to reassess your primary goals.

It is quite obvious that measurement is playing two roles: to monitor the improvement actions and to support deeper analysis of risks associated to weaknesses observed through the assessment phase. Most of the time, the assessment results do not provide immediate actions to be implemented in order to improve processes: a target profile has to be committed according to business goals and difficulties observed.

The relationships between the four activities are illustrated in the following diagram:

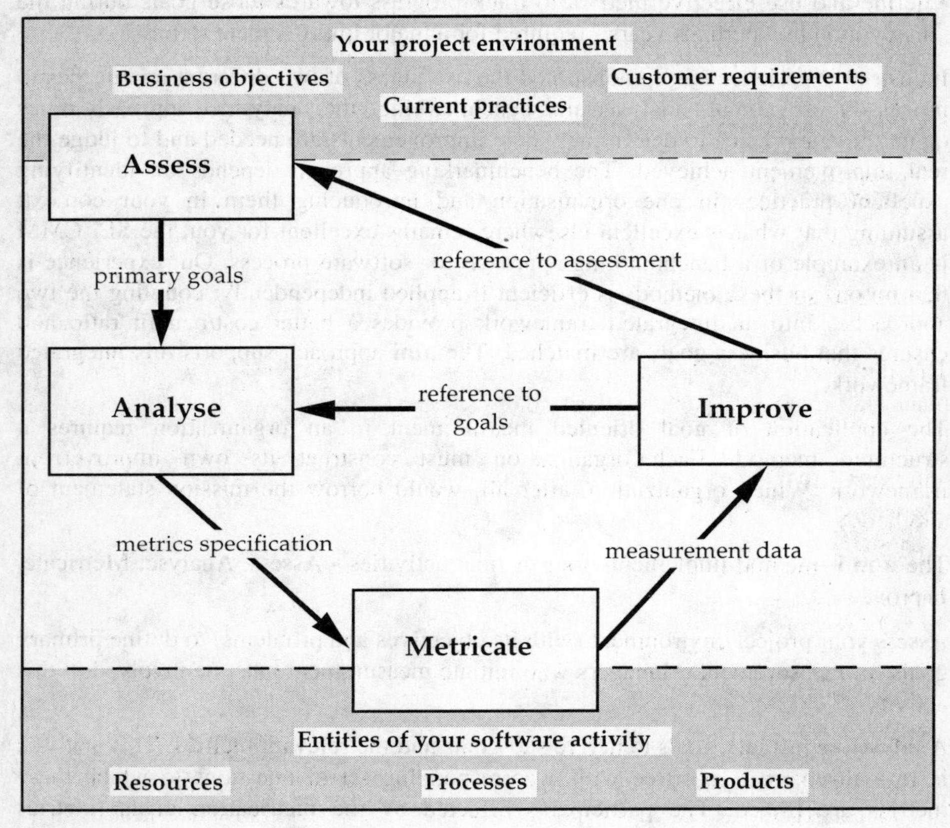

The method is a sequence of 12 steps with a series of support tools (guide-lines, templates and examples) to make it easy to use. ar The first phase of the method addresses how to assess the project environment in order to define and evaluate primary goals for subsequent metrication. These primary goals may be business driven or may evolve more directly from the project environment.

ami steps 1-3

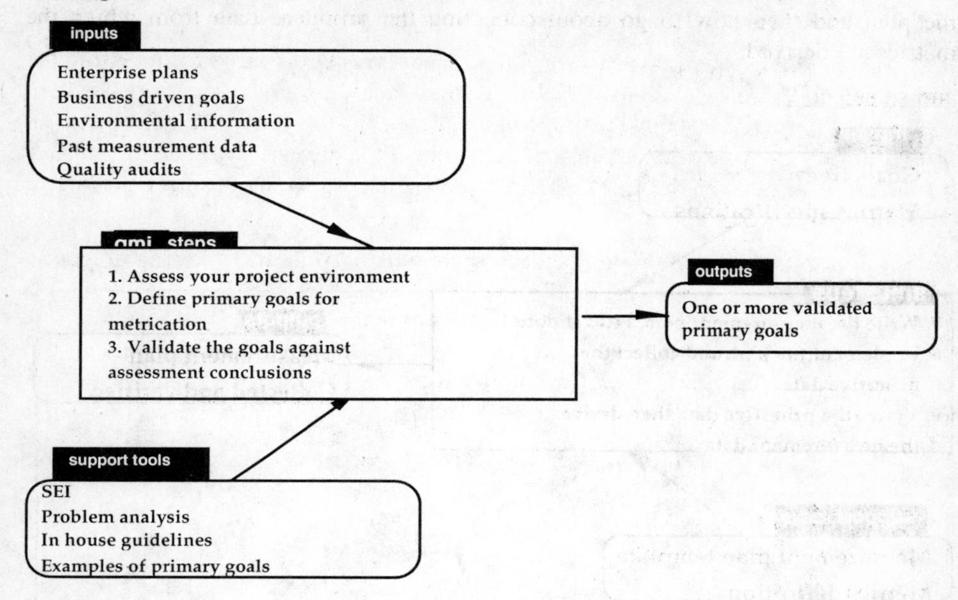

The purpose of the second phase is to break down the primary goals into more manageable sub-goals, and to clarify the measurement objectives in the measurement process.

ami steps 4 to 6

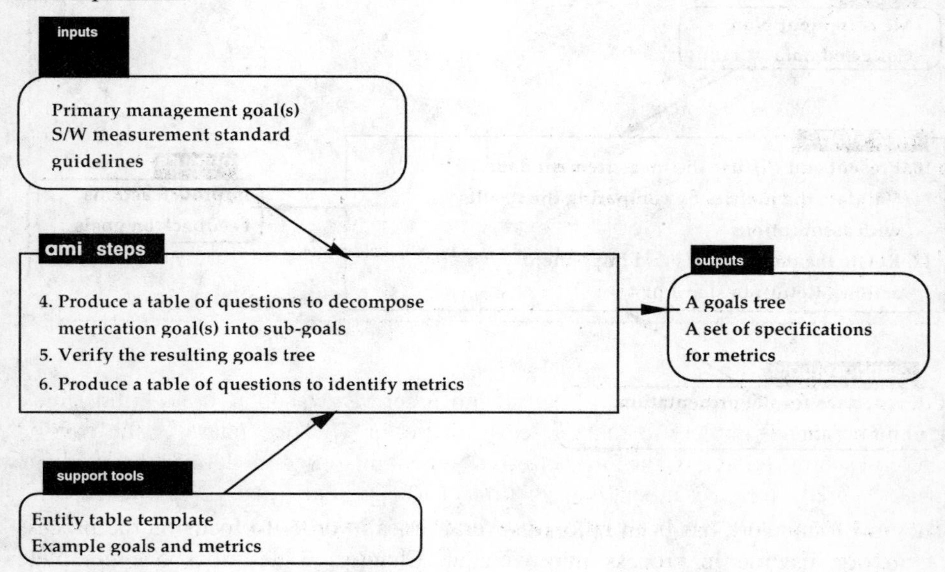

The three next steps describe how to write a measurement plan, how to implement the plan and then how to go about collecting the primitive data from which the metrics are derived.

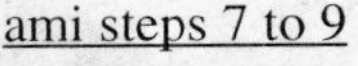
ami steps 7 to 9

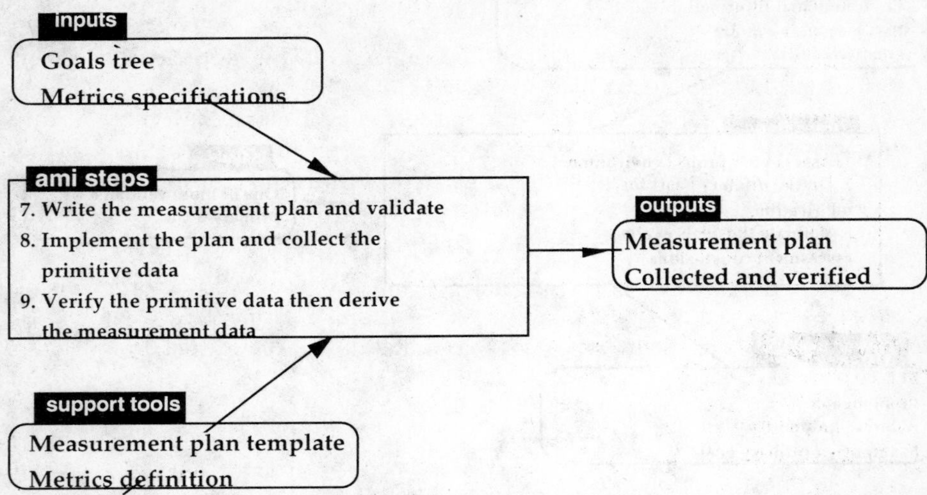

And finally the three last steps of the method make recommendations on how improvement can be achieved through the exploitation of the measurement data.

ami steps 10 to 12

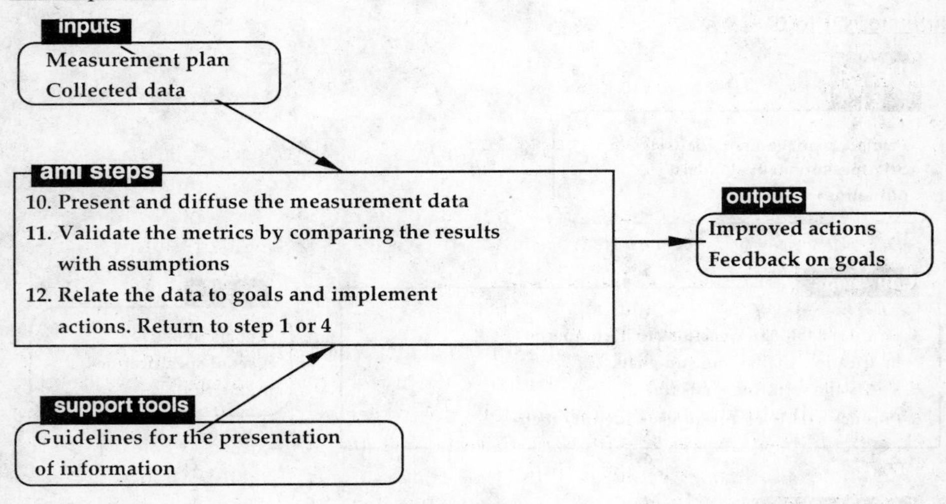

The **ami** framework has been tailored several times in order to focus on the process of an organisation in process improvement. Chapters 3 describes one complete experiences performed in the telecommunication market segment. Chapter 4 tries to clarify the benefits and gives some research trends for improving the efficiency of this type of approach.

3 Experience Description

It was conducted for a group of 300 software engineers involved in the development of various telecommunication network systems. The software part of such systems had been growing by about 20% per year and software development had become the major part of the development investment. At the top management level of this company, a strong commitment has been given to software process improvement; as a result of this willingness, a prior initiative was started during the second quarter of 1993 which OBJECTIF Technologie was requested to support and coordinate.

The initiative was broken down into three main phases: a preparation phase, an assessment phase and an action phase which is still running in 1995. The following paragraphs describe in detail the procedure followed and the lessons learned from this experiment.

3.1 The preparation phase

The work during the preparation phase was concentrated on two main activities : the definition of top management business goals for software improvement and the preparation of the assessment phase.

For achievement of the first activity, several brainstorming sessions were organised based on the general strategy of the company. The **ami** loop served as a road map to the consultant to support this preparation and be sure that the improvement target was achievable in the time and with the initial budget planned.

The role of the consultant at this stage was to coordinate technical with marketing departments, and business strategy with software strategy. This analysis was formalised through several meetings with top technical and marketing managers on the basis of the list of risks observed in the company's life. In front of each main risk a goal for further improvement was defined i.e. to improve the time to market for software development or to introduce software aspects into the sales contracts. These top level goals were then decomposed into strategic sub-goals relevant for technical departments and/or marketing departments.

You may think that his work with top management is unnecessary for several reasons: most of them have no clear idea of the weight of the software development in the overall strategy or have absolutely no experience of software process and difficulties; in other cases, software has become the nightmare for most of the systems and products. But it is vital that the strategic decision to develop and implement a software process improvement supported by a measurement plan is taken by management. Management must also be fully involved in the examination of assessment data and in the definition of primary goals for metrication.

As a consequence, areas of concern and problems in the way in which the organisation currently responds to business and customer needs must first be identified.

In the experiment, the assessment activity was initiated through several meetings with the participants. The software engineering community was first characterised in order to define the number and type of participants. About one third of this community (i.e. 100 engineers) was presumed to participate actively in the assessment. During the preparation phase, the following actions were undertaken:

- customization of the CMM V1.1 model according to the business objectives committed earlier,
- preparation of the questionnaire support: use of the specific company's terminology, definition of questionnaires according to the engineering profiles,
- organisation of awareness sessions for the assessment: explanation of the CMM model, definition of the type of information looked for, presentation of the process to be used for exploiting the information collected among the teams.

3.2 The assessment phase

The assessment was organised over a short period of 3 months in order to get an homogenous picture of the software context. A team of four people performed the various meetings on the basis of the questionnaires and on documents collected on some projects. A total of 150 days was needed to complete the assessment and the preliminary conclusions according to the Key Process Area (KPA) of the CMM procedure.

The main observations reported during this time were the excellent framework and the good cooperation with the software teams. This was mainly due to the initial presentation of the top management goals on one hand and to the preparation meetings organised on the other hand. The majority of the interviewees really wanted to play a role in the improvement initiative and to contribute to the success of the company's business. During the preliminary reviews of the assessment findings and recommendations, fewer than 10% of the observations were discussed or questioned. The global finding of the 100 interviews and discussions (18 project teams) was the following :

- project plans cannot be provided with a high degree of reliability due to missing items such as : recorded past data and information, complete list of tasks including supporting activities or resource allocation flexibility,
- software project tracking and predictability is insufficient,
- difficulties to perform suitable configuration management due to missing definition of work products, activities and absence of tool support,
- software quality activities are generally not performed because of the lack of adequately skilled resources : the role of software quality assurance is not properly defined and corrective actions are usually not taken in time,
- training needs are not always properly addressed, mainly regarding: project management, software quality assurance and software engineering technologies.

Following the detailed analysis of possible consequences of these findings, a list of five main recommendations has been established for the company:

- to consider project experiences as a real company asset; to organise and manage a common database in order to improve project estimates and oversight,
- to identify a minimum set of software quality assurance activities necessary to get the requested level of product quality; to clearly define boundaries between testing and software quality assurance responsibilities,
- to refine the existing software process model for the company and provide guidelines to adapt it to various customers framework,
- to improve testing strategies and tools,
- to install methods and tools for supporting the software configuration management.

From these results, a detailed management plan was prepared for the action phase: customization of these 5 objectives to the different software departments, estimation of resources and schedule for each individual action. In addition, complementary brainstorming was organised at the marketing level in order to identify possible actions that would contribute to the overall success among sales people. The total budget agreed for the first year of the programme was 2% of the company's turnover.

Furthermore, the consultant helped the software improvement programme manager to validate the 5 recommendations against the business goals selected at the end of the preliminary phase dedicated to improvement opportunities identification. This validation resulted in a hierarchy between the recommendations for actions and some adjustment of the quantitative figures associated with the short term business goals.

3.3 The action phase

This phase is still running and the benefits have not yet been completely calculated. The results will be more precise and useful by the end of 1995 when a new assessment will be run. The **ami** approach: Analyse, Metricate, Improve has been selected to support the action plans preparation.

In order to prepare efficient plans, the improvement goals identified by the end of the assessment phase had to be clarified and completed. The company had not chosen to define a target profile - according to CMM model - and to implement the necessary actions to achieve this profile based on a gap analysis between the actual profile and the target. The policy adopted was to rely on the five recommendations coming out from the assessment reports and supposed to help the software teams to remove risks as much as possible. Two options were offered at this stage : either to continue with a complete analytical strategy and install metrics in order to know exactly the origin of the difficulties observed within the teams or to combine analytical analysis and benchmarking based on the experience of the consultant involved into the improvement initiative. The second solution has been selected based on the following reasons:

- the expectation of getting results sooner if adopting a combined approach,

- identification, during the assessment, of fully satisfied practices which could be exported to other teams of the company; this action could be run immediately,
- existence of some useful metrics in the company whose exploitation could help decision making but where not exploited as such i.e. effort planned and spent in each phase of the product development, the number of anomalies reported from the field.

Each department has been working to prepare a detailed action plan including a measurement plan ; working at the department level has been identified as more efficient because of the lack of homogeneity between product lines observed through the CMM assessment : one third had almost completed a level 2, and the others were between levels 1 and 2. The measurement plan was supposed to get continuous feedback of the impact of the actions on the various processes and on the final product.

In order to give an example of the work achieved to prepare actions plan, we will consider the first finding: unreliable and inaccurate estimates. Two types of metrics were defined:

- adequacy metric: the percenntage of accurate initial estimates
- effectiveness metric: the percentage of projects which ends within 10% of the initial estimate.

The actions plan should identify actions in order to help a department to get 95% of projects within 10% of the initial schedule; the present value was 77% of projects completed within 10% of the initial estimated schedule.

Effectiveness measures were used to quantify the improvement goals and define precise actions. The estimate schedule being the output of the planning process, measuring the percentage of projects having made accurate estimates is an indicator of the adequacy of the planning process to perform estimates. If many projects overrun the estimates - i.e. more than 5% - it indicates that the planning process is not satisfying the needs.

Before implementing any action, three main observations were made based on the following measurements:

- the reliability of initial estimates for each phase of the life cycle and the variances,
- the percentage of requirements changes during each phase of the life cycle,
- the percentage of people time spent on reworking software or on recovering from earlier problems.

The current measurements highlighted that:

- up to 75% of task schedule were not matched
- up to 50% of requirements changes were received after the completion of design phase

- up to 50% of the people time was spent on providing solutions to already known problems.

These data were extremely helpful to derive efficient actions and put priorities on process categories to be improved. Requirements management and project planning were identified as key areas for improvement.

As far as project planning was concerned, the following actions were decided:

- Define a procedure to make estimates based on the experience of the recognized experts of the department.
- Define a set of basic metrics to collect on past projects including data related to:
  - the project characteristics and environment
  - the size of the different components, the level of complexity and the number of change requests received,
  - the number of man days spent in each phase of the life cycle for each component.
- Provide mechanisms to store and retrieve the data.

3.4 Lessons learned

The main lesson we learned in the above experiment was the usefulness of combining measurements and the CMM assessment results. Defining effectiveness metrics helps in understanding the real origin of the weaknesses highlighted by the assessment. Furthermore, we demonstrated that, even if measurements are a recognized practice of level 4, the use of simple metrics is useful in the lower maturity level. The importance of Key Process Area crossing the boundaries of a level has been explained in some papers and is emphasized by the SPICE[1] model which differs from CMM in the fact that maturity levels are characterised by common features and not by base practices.

The other main lessons learned are the following:

- when motivation has been stimulated from the very early stages of such an initiative, people are very sensitive to actions and their plan might be overambitious; the consultant has to remain aware of possible frustration that might arise if action plans and capability to achieve them are inconsistent,
- adequate indicators have to be defined for reporting regularly to the top level management: they want to be informed of progress towards the business goals they identified for the programme. In our case we had to define an indicator of return on investment from the beginning of this phase: we selected the ratio effort spent in reviewing and validating a specific phase/effort spent to correct errors introduced at this phase. The ratio should be >1,

---

[1] SPICE: Software Process Improvement and Capability dEtermination, an ISO initiative

- action plans have to be revised regularly to take into account new observations during the improvement project,
- regular reviews have to be organised in order to verify that the effort to fulfil actions is continuous and that the whole team understands the actions.

The first very early result corresponds to an improvement in the homogeneity of the software community capability. Real exchange of know-how, procedures, techniques and tools was observed. This was achieved through the nomination of a software coordinator at the head of each department and through the organisation of a monthly meeting between all coordinators. These coordinators were recognised as software champions by the teams ; all of them were experienced project managers. Now, when there is a need for a specific tool, the decision and selection of the tool is taken by the group of coordinators. This mechanism is a pragmatic way of ensuring reuse of experiences. The next step could be to organise reuse of software between the departments.

The next result concerns the goal "to improve projects estimates and progress". A general procedure has been installed to collect on a monthly basis:

- the effort spent by each person on the tasks he is involved in,
- the necessary remaining effort to complete the tasks,
- the effort spent on non planned tasks : meetings, illness, maintenance of an old product,
- the effort spent on correction of errors introduced in phases that have already been completed (after the corresponding review).

The following indicators have been calculated for each department by the project management tool:

- percentage of unavailability per person for a department; 15% has been observed in almost all departments.
- the average slippage between estimates and reality per category of task for a department; this information is correlated to the progress of the task, i.e. after 50% of task completion or after 70% of task completion,
- the rework ratio: effort spent for correction of errors introduced in a phase/effort spent to develop the phase.

3.5 Observations

Six months after the beginning of the actions, we noticed a change in the behaviour of most of the engineers : individual responsibility resulting from the software improvement initiative was increased. As far as project planning and tracking were concerned, collecting the data became simpler; each engineer validated his own information before providing them to the project manager. Some more advanced departments defined additional metrics to collect in order to improve their performance; they applied the **ami** method, starting form the initial goal "to improve the reliability of the estimates" or "to improve the quality of the final product".

Another observation made is related to the resistance to change among middle managers; they are the real "gatekeepers" of any software process initiative. Sometimes they will encourage and sponsor and sometimes they will impede any progress. In our experiment, we observed both behaviours depending on the department where the initiative was planned. Managers of departments with a higher level of maturity at the beginning were generally more encouraging and tried to keep the priorities among activities in order to sustain actions and key roles. On the other hand, managers of departments where many weaknesses were assessed, were more reluctant to any change. They were not encouraging key people to participate to the working groups set to define actions or allocated part time people (with a low percentage of time).

In such cases, the quantitative observations made were crucial to make the middle managers move. Understanding that rework and/or effort spent to correct bugs that were reoccurring several times consumed a lot of resources made them more flexible to accept a budget line for improvement actions.

The evolution of these figures - rework or bugs correction - when improvement actions are progressively in place makes them enforce the new procedures or processes.

Obviously the fact that the **ami** approach is both pragmatic and based on business goals and measurement goals has been a positive factor in this experience. From the top level management to the software engineer, the method gave better confidence in the return on investment of the improvement programme. Top level management or middle management were satisfied with receiving quantitative feedback from the projects on a regular basis; they were able to follow the impact of software on the business lines and furthermore to go deeper into the analysis of difficulties or weaknesses. The team engineers were quite satisfied with understanding how they would contribute to the overall strategy; in addition, this promoted improvement in their daily work and/or in the procedures.

4 Conclusions

The experience described above illustrates the flexibility of the **ami** method and its effectiveness in defining usable metrics. Positive results of the application of **ami** are reported regularly by the sites where the method has been introduced with a real improvement policy.

The **ami** User Group was set up in early 1993 with support from the department of Trade and Industry and the European Commission through its VALUE[1]'s programme. The overall objective of the User Group is to continue to promote and support the uptake and use of software metrics, and in particular the **ami** method, both within and outside Europe. It provides a range of products and services including the **ami** Handbook, quarterly editions of the **ami** newsletter ***DE FACTO***... **ami**Tool. Members of the User Group have ensured active promotion of **ami** particularly in the UK, Italy and France where a number of seminars and workshops have been organised. The handbook itself has been distributed to over 1000 sites throughout Europe.

We do believe that software measurement should be integrated into the software development process as a critical activity. Actions derived from a metrication initiative are usually cost effective; if you understand the software process, you will identify the critical activities and how to get data about them. The next step should be improving the process through an optimisation of the effort to achieve a given level of quality for the final product. Metrication is iterative - the more you collect, the better you know how to collect, what to collect and how to act from this point.

Measurement that is used for both the control and improvement of software projects gives the following benefits:

Planning, managing and monitoring projects

Measurement enables increased visibility of the quality and progress of a project. Use of data improves predictions and evaluations. Goal orientation improves project coordination. Evolution of real effort with time against predictions is a good indicator of unstable process if a continuous slippage is observed.

Matching the software development process to business objectives

Better matching of the software development process to business objectives can lead to improved confidence between the developer and the customer, between business managers and project managers and between the Software Development Department and the rest of the organization. Communication between people requires quantitative information.

Implementing quality and productivity improvement programmes

The problem of managing investment in software engineering activities is becoming increasingly important. Such investment may be in "preventive" activities such as use of methods, quality assurance, increased training and more testing, or in

---

[1]VALUE stands for Valorisation and Utilisation for Europe. This programme is concerned with the dissemination and utilisation of the results of Community scientific and technological research under programmes such as ESPRIT.

"productive" activities enabled by CASE tools and new technologies. Measurement helps to justify, manage and evaluate such improvement programmes. The production unit -whatever it is- has to be carefully defined before any measurement.

Aiding sub-contracting

Installing a measurement programme to assess the work achieved by sub-contractors is one of the many perspectives of the **ami** approach. Unfortunately, the majority of companies who sub-contract software development on a large-scale rarely differentiate between their capability maturity levels; financial considerations justify the selection. As a consequence, the resulting product does not reach the adequate quality level and/or delay is not matched.

Measurement is not necessary easy. To achieve full benefits, measurement must first be applied systematically according to a customized view of goal-oriented method and be supported by the whole hierarchy. On the other hand, a wide range of developers and engineers should participate and formalise their view of the high level goals for measurement. Very active participation is recommended from both sides; feedback about decisions made after metrics observation and exploitation has to be given to participants. Individual behavior will be changed by metrics initiatives and hostility has to be avoided: metrics do not support individual judgement but collective process.

Acknowledgements

The author express his thanks to the whole **ami** consortium, to the CEC and to the editorial team who have supported the **ami** method development and to the various industrial participants who have contributed to the completeness of **ami** through their positive criticisms. All criticisms received have added clarity and quality to the method.

Bibliography

1. The ami handbook, version 3, December 1992.
2. ami Case studies, Report, January 1993.
3. V.R.Basili and D.Rombach, The TAME project: Towards improvement-orientated software environments, IEEE Trans. Soft. Eng. 14(6), pp.758-773, 1998.
4. B.W.Boehm, Software Risk Management: Principles and Practices, IEEE Software, January, pp 32-41, 1991.
5. D.Card and R.Glass, Measuring Software Design Quality, Prentice Hall, 1990.
6. R.C. Bamford, W.J. Deibler, Comparing, contrasting ISO 9001 and the SEI capability maturity models, IEEE computer, pp 68-70, Oct 1993.

7. C. Debou, J. Liptak, H. Shippers, Decision making for software process improvement: a quantitative approach, In: Proceedings of the 2nd international conference on "achieving quality in software" ACQUIS 93, Venice (Italy), pp 363-377, Oct 1993.

8. N.Fenton, Software metrics: a rigourous approach, Chapman & Hall, 1991.

9. R.B.Grady and D.L.Caswell, Software metrics: Establishing a Company-wide Program, Prentice Hall, 1987.

10. R.B.Grady, Software metrics, Prentice Hall, 1992

11. W.S.Humphreys, Managing the software process, Addison Wesley, 1989.

12. ISO 9000-3: Quality management and quality assurance standards - part 3. Guidelines for the application of ISO 9001 to the development, supply and maintenance of software, International Standards Organization.

13. M. C. Paulk, How ISO 9001 Compares with the CMM, IEEE Software, Jan. 1995.

14. A method for assessing the software Engineering Capability of Contractors, Technical Report, CMU/SEI-87-TR-23.

15. Capability Maturity Model for Software, Tech Report CMU/SEI-93-TR-24, Carnegie-Mellon University, Feb 1993.

BOOTSTRAP: A Software Process Assessment and Improvement Methodology

Pasi Kuvaja
Department of Information Processing Science, University of Oulu
Linnanmaa, FIN-90570 Oulu, FINLAND

Abstract. The BOOTSTRAP methodology for software process assessment and improvement was initially developed by taking the original SEI model as a starting point and extending it with features based on the guidelines from ISO 9000 quality standards and ESA (European Space Agency) process model standards. The extensions were made in order to fit the methodology into the European context, and to attain more detailed capability profiles and maturity levels separately for organisations and their projects. The methodology was also provided with software tools and cumulative assessment data base support. Recently, the methodology has adopted also features of the becoming new ISO standard currently known as a SPICE Initiative. All this has made the BOOTSTRAP methodology particularly suitable for process improvement with different organisation types and sizes and with different kinds of software processes and products. The paper presents the BOOTSTRAP methodology and in conclusions brings forward especially the improvement related aspects that may set boundary conditions between different software process assessment and improvement approaches.

1 Introduction

The value of software process assessment and improvement has already been recognised for some time throughout the software industry. The first step was made by the Software Engineering Institute by publishing its maturity model for software process capability determination in 1987, [15]. Subsequently starting in 1989 an ESPRIT project BOOTSTRAP developed a European process assessment and improvement methodology, where the original SEI model was applied as the main background. Simultaneously the SEI developed new versions of its maturity model now called as Capability Maturity Model (CMM), [11]. Recently an international effort, called Software Process Improvement and Capability dEtermination (SPICE) has been prepared basis for a new ISO standard for the software process improvement and assessment methodologies. The BOOTSTRAP methodology, having participating also in the SPICE Project, [13], is able to address, perhaps, the widest sample of software producing organisations (SPUs) and strongly supports the process improvement activities.

The BOOTSTRAP methodology includes a guided assessment process, maturity and capability determination instruments (questionnaires and algorithm), guidelines for process improvement (standards for action plan generation), and assessor training program. The methodology is supported with computer-based tools and continuously updated European data base that provides an excellent opportunity to compare the maturity and capability levels of the company assessed for example with the rolling means of similar companies. The BOOTSTRAP data base, assessor training and licensing the methodology is taken care by the BOOTSTRAP Institute that was founded by the partners of the BOOTSTRAP Project.

The features that make the BOOTSTRAP methodology in particular suitable for the process improvement are:

- to keep organisation's goals and business needs as a starting point of the process assessment and improvement,
- to use assessment as the first step of the process improvement,
- to use supportive data base and software tools,
- to express capability as detailed profiles,
- to compare organisation and its projects with each other,
- to produce gap analysis profiles for example towards ISO 9001 and SPICE, and
- to maintain and utilise a comprehensive assessment data base as continuously updated industrial benchmark data source.

2 Background models and standards

The BOOTSTRAP assessment and improvement methodology was created within an ESPRIT project of the same name (project number 5441), especially focused onto the European software industry. BOOTSTRAP took into account a selection of methodologies and software process types and especially international software standards applied in Europe.

2.1 The SEI model

Already at the beginning of the BOOTSTRAP project it became clear that efforts to improve software production capability in the industrial companies should be started by assessing the capability levels of the organisation and methodology in software development and maintenance before investing on the process upgrade. Thus, the hypothesis aligned the ideas presented by W. Humphrey ([4]) and it became obvious to adopt the SEI model (later called Capability Maturity Model - CMM, [11], [12]) as a starting point of the BOOTSTRAP approach.

The BOOTSTRAP methodology adopted the maturity scale compatible with the CMM of the SEI. The scale includes five capability stages known as maturity levels: initial level, repeatable level, defined level, managed level, and optimising level. In principle, when a Software Producing Unit (SPU) undergoes improvement, it passes through the levels in this order, and its capability is determined as the last satisfied maturity level. The BOOTSTRAP methodology enhanced the traditional SEI maturity levels by dividing each of them into four quartiles and started to measure the process capability using this quartile scale accordingly within each of the five maturity levels.

2.2 ISO 9000 standards

At the early phase of the BOOTSTRAP methodology development, ISO 9001 was considered the emerging standard in the European market, and it was becoming a mandatory requirement for companies intending to participate international tenders. ISO 9001 is a general quality system standard addressing every kind of companies but additionally provides also guidelines for implementation of a software quality system. The guidelines are included in ISO 9000-3 *"Guidelines for the application of ISO 9001 to the development, supply and maintenance of software"*.

In the BOOTSTRAP methodology the standards were taken into account from two main point of views. At first, as ISO 9001 compliance requires that a quality system is a permanent organism of the company, the requirements for a quality system was captured into the methodology. Effects of the adaptation were that the scope of the assessment was enhanced to involve also the organisation of the SPU as a whole and not only the organisation of the assessed project as the case usually has been with assessment approaches conformant to DoD 2167a standard. Secondly, the distinction between life cycle dependent and independent activities was captured and added to the process functions derived based on the SEI model.

Effects of application of the ISO-9000 standards may clearly be noticed in the BOOTSTRAP methodology, for example, so that the BOOTSTRAP assessment is performed at two levels (organisation and project levels) as also ISO-9000 certification scheme assumes. All together the choices facilitated the BOOTSTRAP assessment to be properly used either when a company is going to a more general improvement program or when it considers ISO 9001 certification as the first step of software process improvement. As a matter of fact it is now generally agreed that companies compliant to ISO 9001 for software fulfil the SEI level 2 to 3.

2.3 The ESA process model

The general process model, which was chosen by the BOOTSTRAP Project for a reference model of the methodology, was a classical software life cycle model that was standardised by the European Space Agency (Standard ESA-PSS 005), [3]. This phase-oriented model had a major influence on the structure of the BOOTSTRAP's assessment instrument developed into the form of two questionnaires. The advantage of the process model chosen was that it had been developed as the basis of earlier standards of IEEE and DoD. On the other hand the selection did not mean that the model was seen as an ideal model, but just a reasonable approach for referring any company specific process to a single model.

2.4 TQM principles

A further model used also as the basis of the BOOTSTRAP methodology was general total quality management (TQM) approach. Some of the main issues of TQM approach were taken into account, as: the idea of a continuous improvement, "Plan Do Check Act"- scheme, process-oriented approach in improving customer satisfaction, and productivity and time-to market as the basic quality dimensions.

Additionally, the general "Kaizen-approach", [5], [16], to improvement lead together with the application of ISO 9000 standard into so-called "Organisation-Methodology-Technology"-paradigm, [10], of improvement in the BOOTSTRAP methodology. The Organisation aspect refers to the overall process organisation, in particular to the quality management and alignment of the improvement actions with the business needs and strategic goals of the SPU. The Methodology aspect fills the organisational framework with procedures and data leading to the SPU targets. The Technology aspect tunes the methodology with particular technology and tool support.

3 Software process assessment

The BOOTSTRAP assessment is performed according to described assessment process steps. The assessment applies scoring and scaling principles to determine where an organization and its projects stand (maturity level). Gap analysis can be used to identify strengths and weaknesses (capability profiles) of the organisation and its processes. The computer-based tools and data base support assessment data collection, results generation and comparison of the results between the organisation and its projects and towards the European data base.

3.1 Types of assessment

Depending on who plays the main role in an assessment three types of assessment are distinguished: first-, second-, and third-party assessments:

- First-party assessment or self-assessment refers primarily to a situation where the assessment is performed internally inside the SPU mainly by SPU's own personnel in order to identify the SPU's own software process capability and, perhaps, initiate or confirm its improvement process. First-party assessment can also be performed as guided first-party assessment where external assessors are used as assessment methodology facilitators for the SPU assessment team.
- Second-party assessment is usually called a capability determination, where external assessors are used to perform the assessment in the SPU.
- Third-party assessment or capability determination is performed by an independent third-party organisation in order to verify the SPU's ability to enter contracts or produce software products.

The BOOTSTRAP methodology can be used for all of these types of assessments, but it is recommended that assessments should be performed with assistance from an external assessor, [17]. The experience from the BOOTSTRAP project shows that self-assessment does not have the same effectiveness and reliability as assisted assessments. One reason for this is that an external assessor has a more formal role in the assessment than internal assessors in self-assessment. Therefore, in the assisted assessment the methodology is applied more rigorously and without any personal tendency to influence the results of the evaluation. Another reason is that an external assessor has wide experience, gained when performing assessments in many other companies. Thus, external assistance assures more reliable results than self-assessment. Additionally the license regulations quarantee that only professional and authorised BOOTSTRAP assessors are allowed to perform the assessment.

3.2 Preparation of the assessment

The preliminary steps that must be taken before the main part of the assessment takes place are: training the people to be interviewed, selecting the organisation to be assessed, defining the assessment team, and signing a confidentiality agreement. The purpose of the training is to get commitment by the assessed organisation, to introduce the main concepts and contents of the process assessment and improvement, and to explain the background standards taken into account in the BOOTSTRAP methodology.

Selection of the target SPU and of the projects to be assessed is performed together with representatives of the client organisation and on the basis of general principles of the BOOTSTRAP methodology. People to be interviewed are also selected according to defined principles, and confidentiality agreement is signed to guarantee a full security and protection of the data collected.

3.3 Assessment execution

After preliminary activities the assessment is performed according to the BOOTSTRAP guidelines. The guidelines foresee the following main activities: opening briefing, global site assessment, project assessment, review of assessment results and presentation of final results. Opening briefing is devoted to people who will participate actively in the assessment. The aim is to reach a collaborative approach and to facilitate the following steps.

So called "global site assessment" is consisted of evaluating the declared production processes. This is done by interviewing the key personnel (e. g. software development responsible, quality assurance responsible) and by evaluating the existing documents like quality manual or company procedures.

So called "project assessment" includes evaluation of software processes in practice that are applied in the assessed projects.

Presentation of the final assessment results is aimed to check the results produced by the assessors and to get commitment on potential improvement activities that was planned and will be introduced on the following phases.

The purpose of the two level assessment is to evaluate how far the projects are to be compliant with the organisation procedures and, on the other hand, to evaluate whether the defined procedures are known, considered applicable and effective in practice.

3.4 Questionnaires as assessment instruments

The BOOTSTRAP methodology includes two separate questionnaires that are used as data gathering instruments and that support and guide the conduct of the interviews performed during the data collection. One of the questionnaires is intended for gathering data on the organisation level and the other is focused on the project level [10].

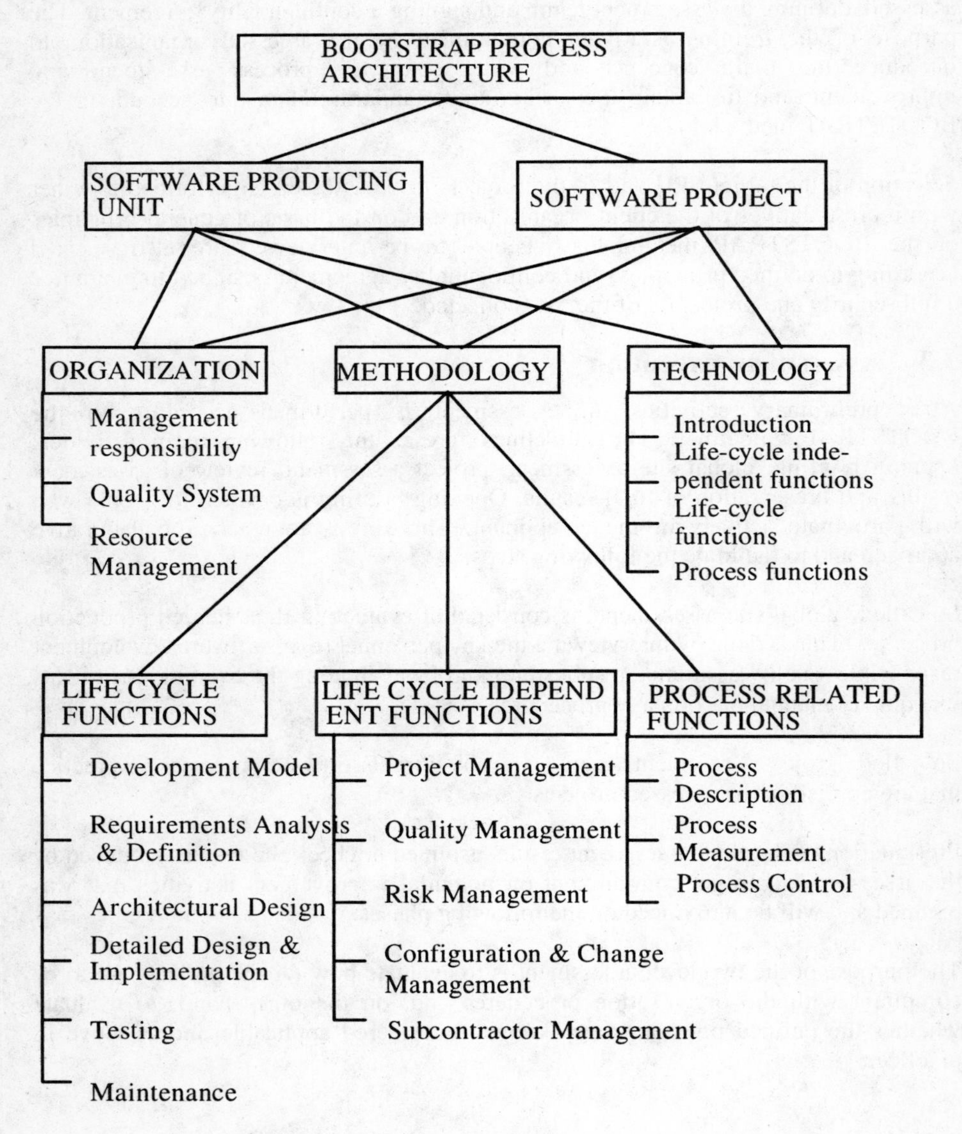

Fig. 1. The BOOTSTRAP process architecture

The questionnaires are used in the guided interviews that are performed as meetings, where the consultants fill the questionnaires on the basis of discussions and documented materials such as quality handbooks, project documentation etc.

The BOOTSTRAP questionnaire is organised as a tree structure which identifies the main attributes of the BOOTSTRAP process model. It is constructed according to process architecture presented above (in Figure 1). The tree - for the purposes of data collection - is implemented as a hierarchical checklist. Individual nodes correspond to hierarchically interrelated sub-checklists. The leafs are derived subsequently into the questions that are individual sensors.

The internal nodes represent synthetic measures, or indicators. The lower down in the tree the less synthetic and more specific an indicator becomes. Indicators higher up in the tree are higher level measures, the root is the top level measure. Higher level indicators give a more global and summary view, lower-level indicators supply more detail; they give a narrower and more specific view of process areas. This applies to both SPUs and projects. The structure allows the following:

- the comparison between different SPUs' quality systems,
- the comparison of two projects from the same or similar SPUs,
- the structure is the basis for the derivation of improvement action plans.

3.5 Scoring principles

The individual questions of the questionnaires are answered by using a score of five values, represented most commonly with such adjectives as: absent, weak, fair, extensive and non-applicable. Thus the difference to CMM is obvious, because the "yes"-answer is in the BOOTSTRAP approach divided into three different intensity levels, [10]. This allows to define more precisely the attribute profiles and helps in capturing the answers of the features that are of various degree of "yes". The questions that cannot be applied in the organisation or project can be left out as non-applicable questions.

Scoring is a sensitive part of the assessment. Repeatability and consistency across different SPUs and projects are the main goals in scoring individual questions. Scoring is determined by the responses to the assessors' questions. Scoring should not be coloured by specific aspects of the context of the SPU or a project. Context dependency is considered later, in the development of an action plan for process improvement. It is also in this later stage that a maturity level for a specific process attribute - corresponding to a sub-checklist is considered.

In order to quarantee as objective scores as possible the recommended scoring procedure to be followed during the assessment session (see Figure 2) is:

- phase I: analysis (during the assessment meeting),
- phase II: clarification (during the assessment meeting),

- phase III: recapitulation (after the assessment meeting, but before the scores are communicated to the site professionals).

During this three-phased scoring process is recommended for assessors to carefully justify the scores and describe their justifications in the space allowed under "Notes" in the questionnaire.

Another factor that aim to get as objective assessment results as possible is to use two authorised assessors in each assessment session. Responsibilities are recommended to be divided between these assessors so that one assessor conducts the discussions while the other records the assessment data using the session tool. In this way it also possible to capture at this first scoring phase potentially different initial scores given by the assessors and come into consensus with them later in the scoring phase three, after that only one score is subsequent stored into the data base and used as a basis of results generation.

3.6 Scaling algorithm

Determining the maturity level from the assessment results should be objective, and should always follow the same rules. For these reasons the BOOTSTRAP algorithm was defined to quarantee compatibility of the maturity levels with the SEI model and to produce a capability profile that express the strengths and weaknesses of the assessed SPU or project. The SEI's maturity model (CMM) was applied in the algorithm as the overall five-level scale that subsequently was made more precise by determining capability using quartile precision inside each maturity level. Therefore, the lowest possible maturity level is 1 and the highest level 5 quartile 4. The algorithm also takes care of the three intensity levels of yes.

The capability profile shows the maturity levels of each process and sub-process called as attributes. Through the capability profiles it can be seen which attributes decrease the total maturity level and therefore they show the most promising improvement areas.

The BOOTSTRAP algorithm is also able to focus the calculation onto selected individual attributes, yielding data profiles that meet organisation's needs and business goals and therefore are useful in identifying the problem areas in the organisational context. This feature can be used with both the SPUs and the projects. It is also possible to produce comparisons between SPU and its projects([10]).

3.7 Assessment results

The following outputs are recommended to be produced for the SPU capability reports:

- a SPU maturity tree presenting hierarchical description of the maturity of the SPU on different levels of the BOOTSTRAP process architecture,
- current SPU profiles presenting current capability of organization, methodology and technology as overall and detailed descriptions,
- SPU strength and weaknesses profiles comparing each process attributes against the SPU overall mean,

- target SPU profiles describing desired capability levels in graphical form,
- gap SPU profiles comparing each process attributes against the chosen targets that may be for example ISO 9001 profile, European mean value, SPICE profile etc..

The following outputs are recommended to be produced for the project capability reports:

- project maturity tree presenting hierarchical description of the maturity of the project on different levels of the BOOTSTRAP process architecture,
- project profiles presenting capability of organization, methodology and technology of the project as overall and detailed descriptions,
- project strength and weaknesses profiles comparing each process attributes against the project overall mean,
- gap profiles comparing each process attributes against the SPU attributes, European mean value, SPICE levels etc..

3.8 Assessment tools and database

The supportive software tools of the BOOTSTRAP methodology include the following main components:

- the session tool,
- the reporting tool,
- database with database management tool.

The assessing tool supports the individual activities of the assessors and the conduct of the BOOTSTRAP assessment session as a whole. The reporting tool supports the production of various analyses of the assessment results and of summary analyses. The database collects, maintains, and provides access to the assessment data. The database management tool is used by the technical personnel of the BOOTSTRAP Institute to monitor and control the activity of the database. Both assessment and reporting tools may be used remotely from the database which supports their activity.

4 Software process improvement

The BOOTSTRAP methodology for software process assessment and improvement may help organisations that are willing to improve their software process. The following three examples describe typical cases, [10], where the BOOTSTRAP methodology may be useful:

- 3K, a software house, had successfully passed their ISO 9001 certification. The managers needed a methodology to create a high-granularity improvement program for fine tuning of the company's software process ...
- KINO Cellular, a large communications manufacturer, that uses software as a competitive weapon in improving their products, needs a software process that can continuously face the high quality demands of the markets and that is cost-effective and able to produce the quality software efficiently. Specifically KINO Cellular needs a refined approach, less general than ISO 9001, suitable for their highly specialised embedded software process, an approach that guarantees continuous improvements ...

- AMEL, a specialised software developer situated in a rapidly developing area, was undergoing a major business reconstruction where they wanted to use local subcontractors. Therefore they wanted to assess the capability of their potential suppliers as they themselves have time to time been assessed by their own main contractors. The assessments had double significance for AMEL - to choose the best suppliers and to help the suppliers to improve their processes for better quality purposes ...

4.1 Improvement principles

TQM, Total Quality Management is a philosophy emphasising continuous organization improvement in order to enhance its competitive position and meet the customer expectations through permanently improved products. TQM emphasises customer focus, management responsibility, training, and employee involvement. The recommendations of the BOOTSTRAP methodology are close if not identical with those of TQM. In practice repetitive assessments with continuous improvement process are concrete results of these aspects in the methodology.

The Kaizen/Shewhart cyclic approach, [5], [16], originated from the need for a strategy that achieves a result through planning and analysis of the response to small increments towards the goal. The method applies many iterations of a cycle of the same structure, including the steps of Plan - Do - Check - Act. The method may optionally include a start-up Initiate stage that creates a framework necessary for effective implementation of the iterations.
The application of the ISO 9000 aspects in the TQM context within the BOOTSTRAP methodology requires that the improvement steps cover related technology aspects in this, [10], order:

- Organization.
- Methodology.
- Technology.

Most software process improvement frameworks include staging recommendations known as the maturity scale. Their role is to bound opportunistic improvements in favour of a long-range improvement program. The BOOTSTRAP methodology also recommends the use of a five-level maturity scale compatible with the CMM of the SEI. Important enhancements to the traditional maturity level philosophy of process improvement in the BOOTSTRAP methodology are as follows:

- capability may be defined also between maturity levels on the basis of a quartile scale dividing each maturity level into four quartiles,
- the capability results may be focused on pre-selected major functions of the process areas and expressed in details with certain process attributes,
- the capability is determined both for the SPU and its projects, which then can be compared with one another and the European rolling means taken from the BOOTSTRAP database.
- the capability profiles may also be focused to express the strength and weaknesses of an SPU and the projects.

4.2 Process improvement steps

The BOOTSTRAP methodology has been already from its beginning aligned with the basic ideas of the Process Improvement Guide of the SPICE project. Therefore, the SPICE terminology is applied here to explain the improvement steps assumed to be taken in the BOOTSTRAP process improvement.

First of all the BOOTSTRAP methodology presumes that organisation's needs and business goals should define starting points of the process improvement and form the primary basis for identifying improvement actions and their priorities. Secondly, the methodology keeps assessment as obligatory first step in the process improvement because it provides a comprehensive picture of the status of the SPU and its projects. The methodology recommends also to use additional effectiveness measures together with target capability in setting the improvement objectives and controlling their attainment together with potential re-assessments. It is also assumed that improvements should be accomplished by undertaking a number of improvement actions, which gradually introduce new and more adequate organisational practices, methodology or technology into software processes or removing old ones.

The BOOTSTRAP methodology assumes that the process improvement whenever started should be considered as a continuous process in the organisation. In practice continuous improvement may be organised as a sequence of steps that form an unending cycle as described below. The main steps of the BOOTSTRAP improvement model are:

- examine organisation's needs,
- initiate process improvement,
- prepare and conduct process assessment,
- analyse assessment results and derive action plan,
- implement improvement, and
- confirm the improvement action completion.

Therefore, it is obvious that in the BOOTSTRAP methodology the main emphasis is on the initiation steps of the process improvement, performance of an assessment and formulate the improvement action plans based on the assessment results. In addition to the assessment focus, the steps of the BOOTSTRAP methodology cover two inner cycles of the improvement model presented below (see Figure 2). These cycles contain the steps of action plan generation, improvement implementation, and improvement action completion validation.

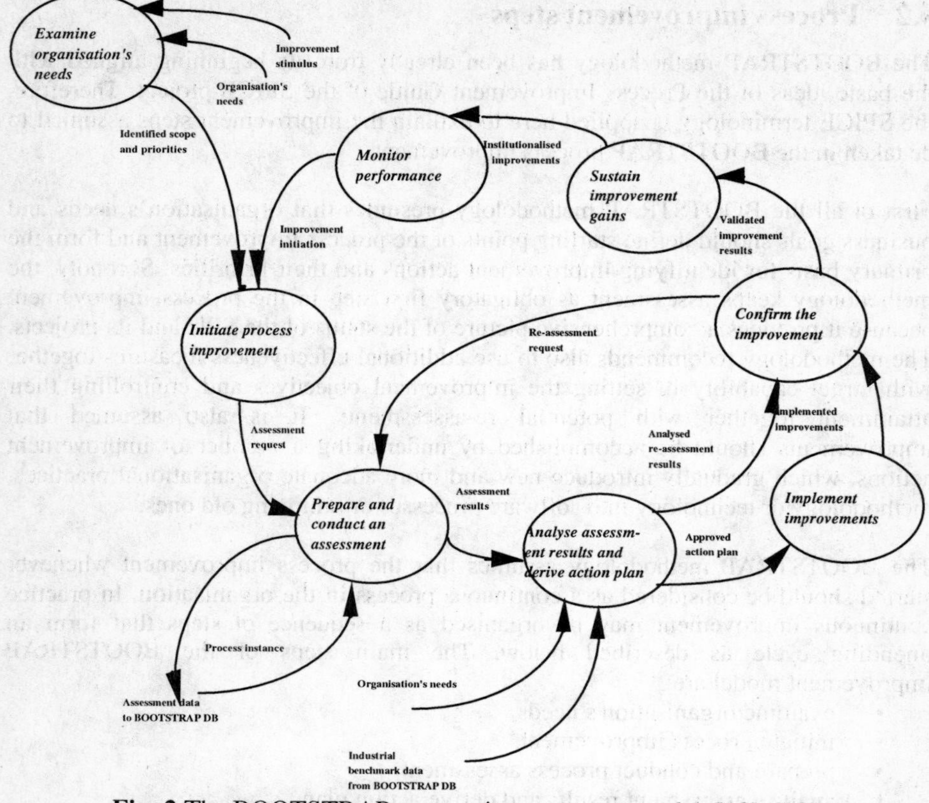

Fig. 2. The BOOTSTRAP process improvement model, [18], [19]

4.3 Selection of the improvement actions

There are two general approaches to select improvement actions, namely analytical and best-practice - driven. The analytical approach attempts to provide detailed descriptions of the improvement solutions. The approach potentially allows for fine optimisation and tuning of the solutions. A negative aspect of the approach is the complexity of the descriptions and the resulting complexity of the planning and designing process.

In the BOOTSTRAP methodology the second one was chosen with some elements of the first. The best-practice approach seeks for practices that have proven to be successful in reaching process improvement goals in some SPUs, and recommends them for application elsewhere. The best-practice approach avoids the problems of complex solution descriptions that are common to the analytical approach. A problem here is, however, that practices working in one environment, due to complexity and other human factors in the process, do not necessarily work in another environment. The discipline of technology transfer attempts to solve this problem.

The BOOTSTRAP methodology recommends the use of best practices corresponding to the indications included in its process model adopted from the ESA. It goes in line with other process assessment and improvement initiatives, among which some give additional guidance about the best practices recommended such as ISO 9001/9000-3 or the SEI's CMM. They are, however, generally non prescriptive in the sense that they give advice about the ends but not the means. The ISO 9004-4 standard [8] may be used as a source for general guidelines referring to quality improvement, although not specifically related to software processes. (Currently developed, the SPICE [14] Baseline Practices Guide - BPG, [20], attempts to support specifically software processes). Moreover, the order of Organisation-Methodology-Technology decomposition of the solutions is assumed to be applied with the improvement action selection in the BOOTSTRAP methodology.

4.4 Action plan generation

The objective of the BOOTSTRAP software process improvement is to assist organisations to start a continuous software process improvement and support its performance with re-assessment and action plan generation. Action plan generation is a step in the whole improvement cycle that contains seven sub-steps as follows:

- check current status,
- provide detailed analysis,
- derive target profiles,
- identify improvement areas,
- prioritise improvement areas,
- derive improvement actions plans, and
- prepare report.

When checking the current status the information collected during the assessment, in particular the current capability profiles, is analysed in the light of the organisation's needs to improvement. The detailed analysis produces profiles, where the processes are compared to organisation's own mean capability levels, and against the industrial benchmark data retrieved from the BOOTSTRAP data base. These analyses are also completed with appropriate notes collected during the assessment. Target profiles are then derived based on organisations improvement goals. Identification of the improvement areas will be performed based on the results of a gap analysis between the current status and the target profiles. Typical general gap analysis results are comparisons against the ISO 9001 and SPICE profiles. Prioritising the improvement areas should be based on analysis of the organisation's needs, and risks to the organisation and its products in order to ensure that the subsequent improvement actions will best meet the needs and mitigate the risks. Finally a set of actions to improve software processes of the organisation will be developed. A typical report of the BOOTSTRAP assessment and improvement activity contains elements outlined below.

1. Executive Summary
2. Information on the Site Environment
3. Comments on the SPU current status
    3.1. Detailed Comments on Site Organisation
    3.2. Detailed Comments on Site Methodology
    3.3. Detailed Comments on Site Technology
4. Maturity Levels for the Site
5. Site Action Plan Recommendations
    5.1. Recommended Maturity Levels
    5.2. Recommended Action Plan
        5.2.1. Organisation
        5.2.2. Methodology
        5.2.3. Technology
6. Timetable

Fig. 3. An outline of the SPU assessment report [10].

5 Assessment results and experiences

All the assessment data collected during the BOOTSTRAP assessments are collected into a data base under a high confidentiality. The data base do not contain organisation's name of or any other identification items. Only the assessors know what organisations they have assessed. Therefore, the BOOTSTRAP data base is able to provide information that can be used to support the assessment results comparison against all other data in the data base. Here some summary information will be presented based on the analysis of the BOOTSTRAP data base and experiences collected during the assessments.

5.1 The European status

Results of the assessments performed by the BOOTSTRAP assessors contain overall maturity levels of the SPUs and projects and mean values of the key attributes forming capability profiles of the SPUs and projects. Here only maturity level results are presented. Capability profiles are kept confidential. The maturity levels shown below have been counted based on the assessed 37 SPUs and 90 projects up to the end of year 1993. Figure 4 presents the percentage distribution of SPUs on different maturity level quartiles and figure 5 the same for projects.

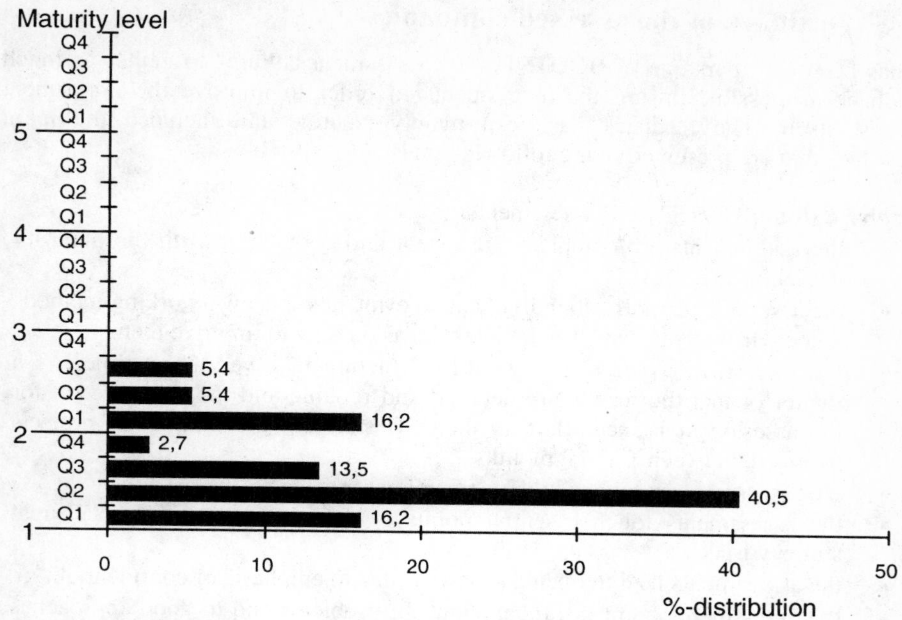

Fig. 4. Mean maturity of SPUs assessed up to 31.12.1993.

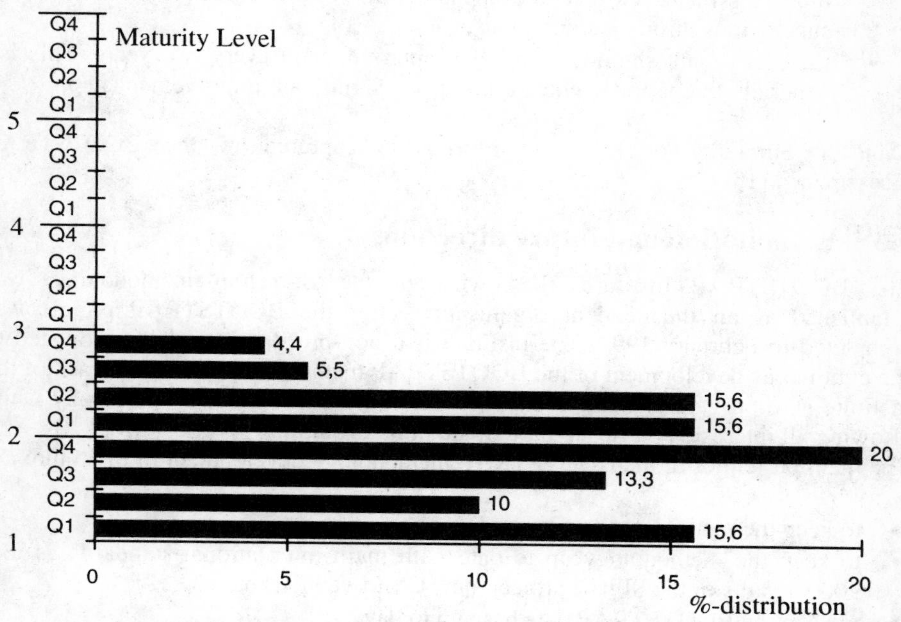

Fig. 5. Mean maturity of projects assessed up to 31.12.1993.

5.2 Feedback of the assessed companies

It has been a cornerstone of BOOTSTRAP "assessment culture" to gather as much feedback as possible during the assessments in order to improve the assessment process itself. The feedback has been mainly positive and included the main messages that are presented in the following table.

Feedback from BOOTSTRAP assessments:

- the assessments are complete, consistent and compatible with the maturity model of SEI,
- the assessments motivate individuals to evaluate their own working methods and environments and stimulate new ideas on how to improve them,
- the assessments give a very good SPU picture in a very short time. As a matter of fact the results are detailed and reliable and the time needed for the assessment is, according to the SPU size and the number of projects assessed, between 1 and 3 months,
- the assessments receive a high degree of acceptance,
- the assessments touch essential points of daily work, as well as critical strategy issues,
- the assessments lead to open discussions due to emphasis of confidentiality,
- the assessments point to fundamental SE problems and to good approaches for improvements,
- the assessments enforce the capability for improvements by constructive suggestions through action plans,
- the assessment should be repeated approximately every two years. This is generally the period needed to perform a significant improvement step.

Table 1 Summary of feedback information captured in the BOOTSTRAP assessments, [10].

5.3 Exploitation and future directions

The BOOTSTRAP Institute, the owner of the Bootstrap methodology, was established as an independent organisation after the BOOTSTRAP project was completed in February 1993. The institute is a non-profit organisation dedicated to the continuous development of the BOOTSTRAP methodology. The main task of the institute is to provide fair and equal access to the BOOTSTRAP methodology while allowing all interested parties to participate in its evolution.
The main principles of the BOOTSTRAP methodology development in the future are as follows:

- to keep the methodology consistent,
- to keep the methodology up to date with main international standards as ISO 9000, results of the SPICE project, new CMM versions etc.,
- to keep the BOOTSTRAP data base up to date,
- to maintain the supportive computer-based tools,

- to ensure that all BOOTSTRAP assessments are performed according the BOOTSTRAP principles,
- to control the use of the methodology, following the terms of the license agreements,
- to guarantee that the methodology keeps its position as the leading European software process assessment and improvement methodology in the markets.

In order to fulfil its mission in the future the BOOTSTRAP Institute will continue close participation in the ISO software quality activities and co-operation with the European Software Institute.

6 Conclusions

BOOTSTRAP has got an established position in Europe. It is important for the methodology to look for comparisons of assumptions and results elsewhere. Such comparisons have been made world widely during the SPICE project and the current version of the BOOTSTRAP methodology has already adopted new features from the SPICE work. Important aspects such as the underlying process model, the process scaling procedures, process improvement model, and the technology transfer issues will become mandatory to all assessment and improvement methodologies in the world when the results of the SPICE project will form the new ISO standard. The first two aspects have already been taken into account in the development of the current version of the BOOTSTRAP methodology. The last two aspects - process improvement and technology transfer - are still under active development in the methodology. It is believed that a wider exposure to the international community, experienced currently by BOOTSTRAP, will facilitate these investigations.

References

1. Kuvaja, P., and Bicego, A., BOOTSTRAP: Europe's assessment method, IEEE Software, Vol. 10, Nr. 3 (May 1993), pp. 93-95.
2. Military Standard. Defence System Software Development. DOD-STD-2167A (29 February 1988).
3. ESA Software Engineering Standards ESA PSS-05-0. Issue 2. ESA Board for Software Standardisation and Control, European Space Agency, Paris (February 1991).
4. Humphrey, W. S., Managing the software process. Addison-Wesley Publishing Company Inc., Reading, Mass. (1989).
5. Imai, M., Kaizen: The Key to Japan's Competitive Success. Random House, New York (1986).
6. ISO 9000-3. Quality management and quality assurance standards. International Standard. Part 3: Guidelines for the Application of ISO 9001 to the Development, Supply and Maintenance of Software. ISO (1991).

7. ISO 9001. Quality Systems. Model for Quality Assurance in Design/Development, Production, Installation and Servicing. International Organisation for Standardisation, Geneva (1989)._
8. ISO 9004-4. Quality management and quality system elements. International Standard. Part 4: Guidelines for quality improvement. ISO (1993).
9. ISO 9126. Information technology - Software product evaluation - Quality characteristics and guidelines for their use. International Standard. ISO (1990).
10. Kuvaja, P., Similä, J., Krzanik, L., Bicego, A., Koch, G., and Saukkonen, S., Software Process Assessment and Improvement. The BOOTSTRAP Approach. Blackwell Business, Oxford, UK, and Cambridge, MA 1994.
11. Paulk, M., et al. Capability Maturity Model for Software, Version 1.1, CMU/SEI-93-TR-24, Feb. 1993.
12. Paulk, M., et al. Key Practices of the Capability Maturity Model, Version 1.1, CMU/SEI-93-TR-25, Feb. 1993.
13. Paulk, M.C., and Konrad, M.D., An overview of ISO's SPICE project. American Programmer (February 1994).
14. SPICE - Software process capability determination standard product specification for a software process capability determination standard. Document WG10/N016. ISO/IES JTC1/SC7/WG10 (1993)._
15. Humphrey, W.S. and Sweet, W.L., "A Method for Assessing the Software Engineering Capability of Contractors", SEI Technical Report SEI-87-TR-23, September 1987.
16. Huda F. and Preston D., "Kaizen": The Applicability of Japanese Techniques to IT, Software Quality Journal, No.1, pp. 9 -26, 1992, Chapman Hall.
17. Bicego A., The SEI Model and the BOOTSTRAP Methodology, in the Proceedings of the Esprit Scope - BOOTSTRAP Conference on Measurement for the Software Industry Today, 17 - 18 September 1992, Milan, ITALY.
18. Kuvaja P. and Bicego A., SPICE Process Improvement Guide, SPICE Tutorial in the Fourth European Conference on Software Quality, Basel, Switzerland, October, 1994.
19. Kuvaja P. , Software Process Assessment and Improvement - The BOOTSTRAP Approach, in the Proceedings of CQS - European Observatory on Software Engineering: Improving Processes, Products and Services, Rome ,Italy, September, 1994.
20. Konrad M., SPICE Baseline Practices Guide, SPICE Tutorial in the Fourth European Conference on Software Quality, Basel, Switzerland, October, 1994.

The *SPICE* Project: An International Standard for Software Process Assessment, Improvement and Capability Determination

Antonio Coletta
Tecnopolis CSATA Novus Ortus, Strada Prov. per Casamassima km. 3
70010 Valenzano (BA), Italy, Tel. +39 80 8770256, Fax. +39 80 8770521
Internet : coletta@max.csata.it

Abstract. SPICE (Software Process Improvement and Capability dEtermination) is an international collaborative effort to develop a Standard for Software Process Assessment under the auspices of the International Committee on Software Engineering standard, ISO/IEC JTC1/SC7/WG10. This paper provides an overview of the project and its current results (now undergoing world-wide trials). Some technical details are also given on the Process Capability Model, and on the Measurement Framework which are the two fundamental elements of the standard. The Process Capability Model identifies a set of software best practices and shows how they fit together to create an improvement path for the software development process. The Measurement Framework is used during the process assessment activities to evaluate the implemented practices and generate a Capability Profile for each process assessed.

1 Project Background and Organization

In June 1992 the ISO/IEC Joint Technical Committee 1, through its Software Engineering Subcommittee (SC7), approved a resolution recommending the creation of a new Working Group (WG10) with the task of developing an international standard on Software Process Assessment. In January 1993, ISO/IEC JTC1 adopted the resolution and assigned the task to WG10.

The decision was taken following a one year study period in which a report [1] had been produced indicating that :

- there was international consensus on the need and requirements for a standard for process assessment
- there was international consensus on the need for a rapid route to development and trialling of the standard in order to provide usable output in an acceptable time scale and to ensure the standard fully met the user needs
- there was international commitment to provide resources for a project team coordinated through four technical development centers based in Europe, North America (Canada and USA) and Pacific Rim.

To speed up the work, WG10 decided to carry out the development stage of the new standard through an international project named SPICE (Software Process

Improvement and Capability dEtermination) and established a *modus operandi* defining the relationship between SPICE and WG10.

The project was kicked-off at the beginning of 1993 in Dublin and is currently supported by numerous organizations from 16 different countries. The project organization is illustrated in fig. 1.

The work is actively sponsored by the major suppliers of Software Process Assessment methods such as SEI (CMM) and the European Bootstrap Consortium and a liaison has been created with the European Software Institute in Bilbao. The aim is to build a standard based on the best features of existing methods, yet providing an innovative common approach. The results are currently being trialled by the international community before going through the standardization stage.

Given the complexity and size of the project, a Quality Management System has been set up and formal procedures have been defined to cover aspects such as documents development, revisions and approval, configuration management, problem reporting, etc.

Four *technical center managers* are basically responsible for regional resources procurement and allocation, while seven different *product managers* have the responsibility for developing the different parts of the standard. The technical centers are based in :

- UK - Defence Research Agency
- Canada - Bell Sygma
- USA - Software Engineering Institute
- Australia - Griffith University

Documents and messages are exchanged between team members through Internet (email), and (ftp) servers are available containing the configuration library of the project results.

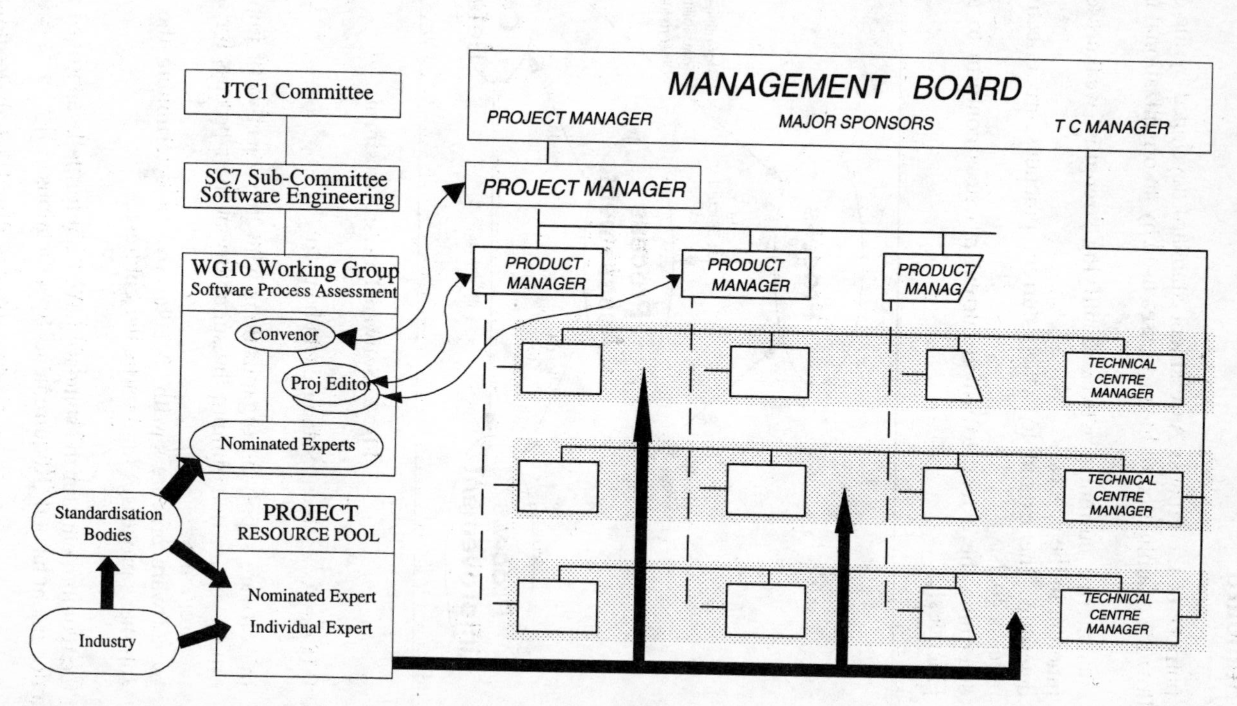

Fig. 1. Project Organization

2 Requirements and Specifications for the Process Assessment Standard

The purpose of the Process Assessment Standard developed by the SPICE project is to allow the examination of the processes used by an organizational unit in order to:

- characterize current practices identifying strengths, weaknesses and the risks inherent in the process,
- determine the extend to which current practices are effective in achieving process goals,
- determine the extend to which current practices conform to a set of baseline practices.

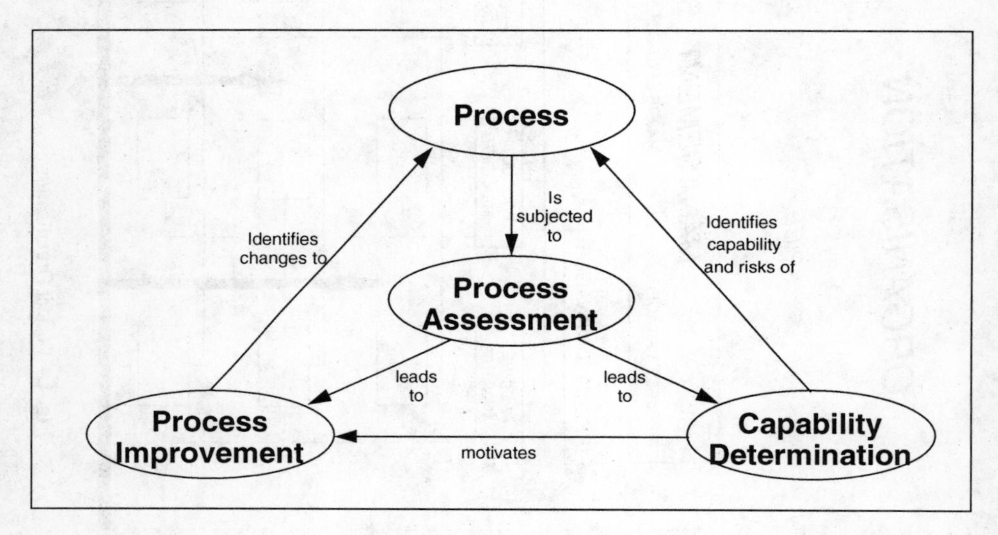

Fig. 2. Software Process Assessment

The Process Assessment Standard can be used:

- by software supplying organizations with the objective of improving their own process or for determining the suitability of their process for a particular set of requirements
- by software procurers with the objective of determining the suitability of the supplier's processes for a particular contract.

All the requirements for the standard have been formally approved and are contained in a document named "Requirements Specifications" [3].

The key functional requirements identified in SPICE are the following:

- the standard must be applicable over a wide range of application domains, businesses and sizes of organization,
- it must be applicable at different levels of organizational unit (projects, departments, corporate, etc..),

- it must define baseline (i.e. essential) practices for each process,
- it must produce an output in terms of process profiles at different levels of detail,
- it must provide comparability between similar entities.

The major non-functional requirements include the following:

- the standard must support reliable and consistent assessments,
- it must be simple to use and understand,
- it must be objective and quantitative wherever possible,
- it must not presume specific organizational structures or management philosophies, lifecycle models, technologies or development methods,
- it must be supportive of existing standards such as ISO 9000 and other SC7 software engineering standards.

The above requirements, once approved by the project team, have been mapped and traced to the "Product Specification" [4] a document containing a high-level specification for each component of the standard (see following section).
Starting from the high level product specifications, each product team has then developed a detailed specification of the component subject to project approval before actual development.

3 The Standard Architecture and Components

The documents developed by the SPICE project, comprising the Process Assessment Standard provide requirements and guidance on how to conduct a *process assessment* using a model of processes and practices essential to good software engineering. They also provide guidance on *Process Improvement*, *Capability Determination*, *Assessors Training and Qualification* and use of *Assessment Instruments*.

The components of the standard and their functional dependencies are illustrated in Fig. 3.

The SPICE suite consists of seven documents:

Part 1: Introductory Guide (IG)

The IG is the top level umbrella document which describes how the other parts of the standard fit together, and provides guidance for their selection and use. It defines criteria for conformance to the standard.

Part 2: Process Improvement Guide (PIG)

The PIG provides guidance on how to prepare for and use the results of an assessment for the purposes of process improvement. It also includes a number of guidance models applicable to particular situations.

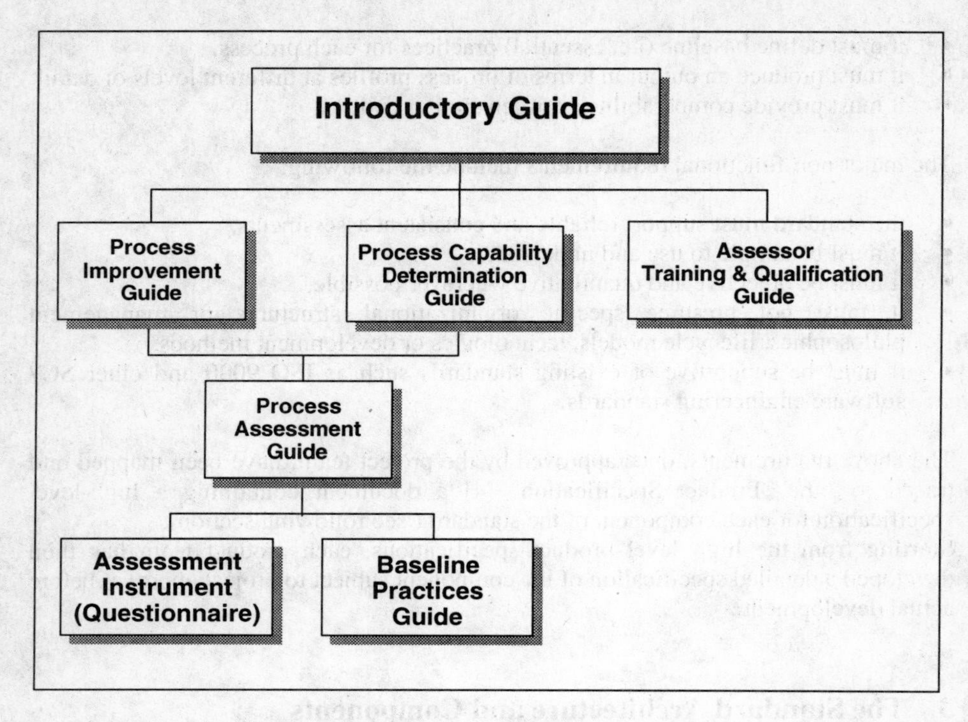

Fig. 3. Components of the SPICE standard

Part 3: Process Capability Determination Guide (PCDG)

The PCDG provides guidance on how to prepare for and use the results of an assessment for the purposes of determining the capability of one or more process within an organizational unit. The guide specifically addresses capability determination within two different scenarios:

(i) for use within an organization to determine its own capability
(ii) by a purchaser to determine the capability of a (potential) supplier

Part 4: Process Assessment Guide (PAG)

The PAG defines how to conduct an assessment, and sets out the basis for rating, scoring and profiling process capabilities. This guidance is generic enough to be applicable across all organizations, and also for assessing in a number of different modes.

Part 5: Baseline Practices Guide (BPG)

The BPG defines, at a high level, the goals and fundamental activities that are essential to software engineering, structured according to increasing levels of process capability. These baseline practices may be extended through the generation of application or sector specific practice guides to take into account specific industry, sector or other requirements.

Part 6: Assessment Instrument (AI)

The AI defines the rules for constructing tools to assist in performing assessments. It provides guidance on indicators for the existence and adequacy of practices. It includes (as Appendices) a number of exemplar assessment instruments.

Part 7: Assessor Training and Qualification Guide (ATQG)

The ATQG standard describes the process of obtaining a qualification for the purpose of carrying out SPICE assessments. ATQG also describes the process of validating a potential assessor's skills, knowledge and attitudes in terms of competencies, or a combination of education, training and experience.

4 The Process Assessment Context

The Process Assessment activity is not an end in itself but is performed either during a process improvement initiative, or as part of a capability determination exercise. As such it is invoked by and returns results to either the PIG or the PCDG.

The context under which a SPICE conformant Process Assessment takes place is illustrated in Fig. 4.

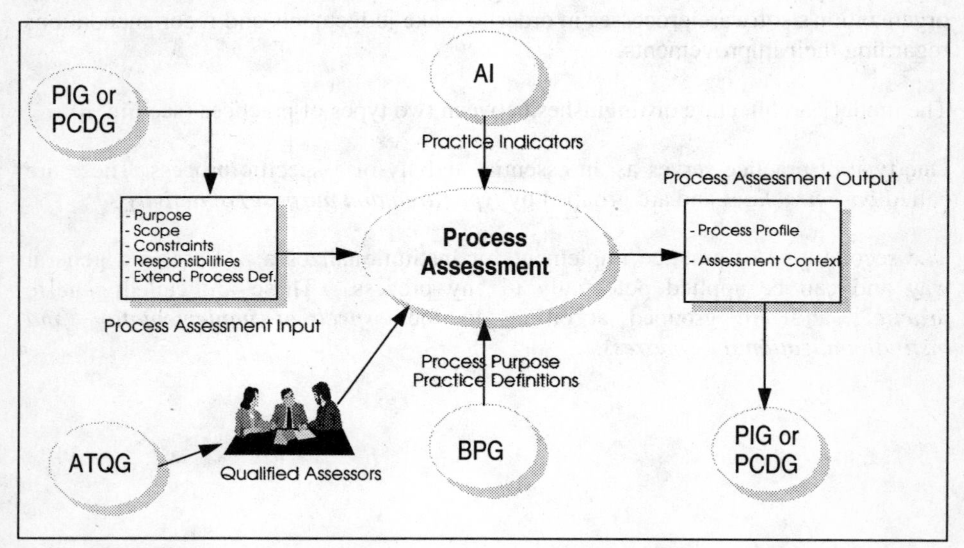

Fig. 4. Process Assessment Context

Whether invoked for Process Improvement or for Capability Determination, there must be an input to the Process Assessment activity defining the purpose (why the assessment is being carried out), scope (what processes should be assessed) and what constraints, if any, apply. The assessment input also defines the responsibility for carrying out the assessment and gives definitions for any processes within the scope of the assessment that are variants of the processes within the Baseline Practices Guide.

The assessment output is returned to the calling process (either process improvement or process capability determination) and consists of the scores assigned to the processes assessed (the process profile), and a record of the context in which these ratings were awarded. The context is important because, among other things, it is used to determine the comparability of different assessment results.

The assessment is conducted by trained and qualified assessors using the Process Capability Model defined in the BPG, the Measurement Framework defined in the PAG and with the support of assessment instruments (typically a questionnaire, a check-list or perhaps an automated tool) conformant to the requirements specified in the AI guide. The Process Capability Model and the Measurement Framework are basic elements of the SPICE approach and are described in more details in the next two sections.

5 The Process Capability Model

The Process Capability Model defined in detail in the Baseline Practices Guide (BPG) identifies practices and processes which should be implemented to establish and improve an organization's software development, maintenance, operation and support capabilities. It has been designed to facilitate the assessment of an organization's software processes in order to make judgements and recommendations regarding their improvements.

The model's architecture distinguishes between two types of practices (see Fig. 5).

One type of practice serves as an essential activity of a specific process. These are called *base practices* and are grouped by *type (area and purpose) of activity*.

The second type of practice implements or institutionalizes a process in a general way and can be applied potentially to any process. These are called *generic practices* and are grouped according to the *aspect of implementation and institutionalization they address*.

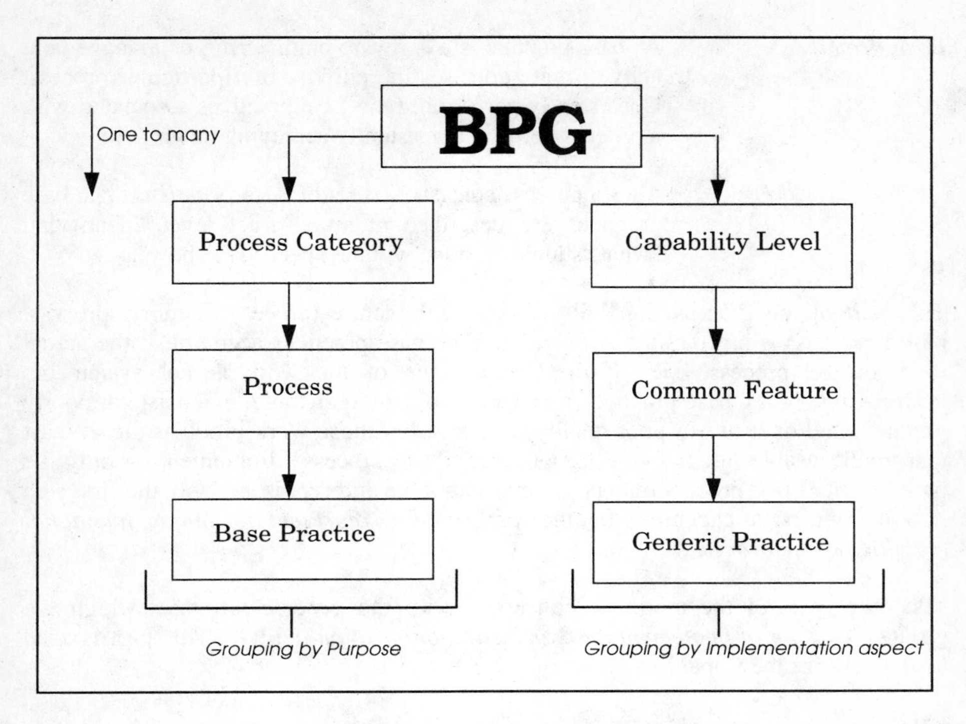

Fig. 5. The Process Capability Model

The *base practices* are hierarchically organized as follows:

Process Category — A process category is a set of processes addressing the same general area of activity. For example, the *Engineering* process category consists of the processes related to the *analysis, design, implementation, review and maintenance* of software.

The process categories covered in the BPG address five general areas of activity: *customer-supplier activity, engineering, project management, support, and organization infrastructure building*.

Process — A process is a set of activities that achieves a purpose. Each process has a purpose and consists of a set of base practices that address that purpose.

The processes as defined by SPICE are not processes in the sense of being complete process models or descriptions. The BPG processes contain a list of essential practices, but they do not describe how to perform the process. The BPG is a descriptive model, not a prescriptive model.

Base Practice

A base practice is a software engineering or management activity that addresses the purpose of a particular process. Consistently performing the base practices associated with a process helps in consistently achieving its purpose.

Thus a process consists of a set of *base practices.* The base practices are described at an abstract level, identifying "what" should be done without specifying "how".

This part of the Process Capability Model architecture (process category, process, base practices) is a grouping by *purpose*. The base practices accomplish the actual work of the process, even if the performance of the work is not systematic. Performance of the base practices may be ad hoc, unpredictable, inconsistent, poorly planned, and/or result in poor quality products, but those work products are at least marginally usable in achieving the purpose of the process. Implementing only the base practices of a process may be of minimal value and represents only the first step in building process capability, but the *base practices represent the unique, functional activities of the process*.

The other part of the model is concerned with the *generic practices* which are grouped by *type of implementation or institutionalization activity*. This part is used to characterize the Capability Level of a process.

Capability Level

A capability level is a set of common features (sets of activities) that work together to provide a major enhancement in the capability to perform a process.

As an example, the *Planned-and-Tracked Level* consists of practices related to planning, managing, and verifying the performance of (any) process.

There are six capability levels in SPICE. Each level provides a major enhancement in capability to that provided by its predecessors in the performance of a process. Together, they also provide a road map for improving a specific process in a logical fashion.

Common Feature

A common feature is a set of practices that address an aspect of process implementation or institutionalization.

As an example, the *Planned-and-Tracked Level* contains the *Planning Performance*, *Disciplined Performance*, *Verifying Performance*, and *Tracking Performance* common features. The *Tracking Performance* common feature consists of practices that track process status using measurement and take corrective action as appropriate.

Generic Practice A generic practice is an implementation or institutionalization practice that enhances the capability to perform any process.
The generic practices apply to a process as a whole and can be aggregated by common features and by capability levels to describe the capability of a process.

This part of the Process Capability Model architecture (capability level, common feature, generic practices) is a grouping by *type of implementation or institutionalization activity*. The generic practices characterize good process management that results in an increasing process capability for any process. A planned, well-defined, measured, and continuously improving process is consistently performed as the generic practices are implemented for a process. This process capability is built on the foundation of the base practices that describe the unique, functional activities of the process.

In summary, Fig. 5 shows graphically all the components of the Capability Model and their relationships.

The linkage between the two sides of the architecture is obtained through the measurement framework during the process assessment activities.

The six Capability Levels defined by SPICE are shown in Fig. 6.

0	**Not-Performed**	There is general failure to perform the base practices in the process. There are no easily identifiable work products or outputs of the process.
1	**Performed-Informally**	Base practices of the process are generally performed but not rigorously planned and tracked. Work products of the process testify to the performance.
2	**Planned-and-Tracked**	Performance of the base practices in the process is planned and tracked. Performance according to specified procedures is verified. Work products conform to specified standards and requirements.
3	**Well-Defined**	Base practices are performed according to a well-defined process using approved, tailored versions of standard, documented processes.
4	**Quantitatively-Controlled**	Detailed measures of performance are collected and analyzed leading to a quantitative understanding of process capability and an improved ability to predict performance.
5	**Continuously-Improving**	Quantitative process effectiveness and efficiency goals (targets) for performance are established, based on the business goals of the organization. Continuous process improvement against these goals is enabled by quantitative feedback.

Fig. 6. The Six Capability Levels

6 The Measurement Framework

The Measurement Framework is the "meter" used during a SPICE assessment to measure the Capability of a process. It is based on the Process Capability Model and its 6 Capability levels.

A process is considered at level 0 when no evidence is available of its existence. If the process exist, then the assessment team will try to understand whether all its base practices are adequately performed (Capability Level 1). From level 2 onward the assessors will try to judge whether the process is managed and institutionalized in terms of being planned, tracked, defined, measured and continuously improved.

The assessor's judgement at each level is formulated using the *Generic Practice Adequacy Rating* defined as *a judgement, within the process context, of the extent to which an implemented Generic Practice satisfies its purpose.*

The scale used for the Adequacy Rating of the Generic Practices has four descrete values:

Not Adequate. The Generic Practice is either not implemented or does not to any degree satisfy its purpose.

Partially Adequate. The implemented Generic Practice does little to satisfy its purpose.

Largely Adequate. The implemented Generic Practice largely satisfies its purpose.

Fully Adequate. The implemented Generic Practice fully satisfies its purpose.

Using this scale, the assessment team examines all process occurrences, within the scope of the assessment, and determines an adequacy rating for each Generic Practice in each Capability Level.

The result, for each Capability Level (except level 0), is a matrix having the format :

CL n = [%N, %P, %L, %F]

where %N = percentage of Generic Practices judged as *not adequate*
%P = percentage of Generic Practices judged as *partially adequate*
%L = percentage of Generic Practices judged as *largely adequate*
%F = percentage of Generic Practices judged as *fully adequate*

and %N + %P + %L + %F = 100 %

One special case is Capability Level 1 (Performed informally), in which there is only one Generic Practice (Perform the Process). In order to evaluate this Generic Practice, it is necessary to look for the existence/adequacy of all the Base Practices belonging to that process and derive a global rating based on how they work together to perform the process. The purpose is to judge the completeness of the process by looking for work products that testify the deployment of each of its Base Practices.

From level 2 upwards, the Generic Practices (more than one per level) are evaluated in relation to the purpose of the entire process and not looking at the single Base Practice.

At this point 5 Capability Level rating matrices are generated for each process by aggregating the Generic Practice Adequacy ratings within each Capability Level (from level 1 to level 5).

A typical output from a SPICE assessment for a process X, may therefore be:

CL1 : [0%, 0%, 40%, 60%]
CL2 : [30%, 35%, 25%, 10%]
CL3 : [80%, 20%, 0%, 0%]
CL4 : [0%, 0%, 0%, 0%]
CL5 : [0%, 0%, 0%, 0%]

which can be shown graphically as in Fig. 7.

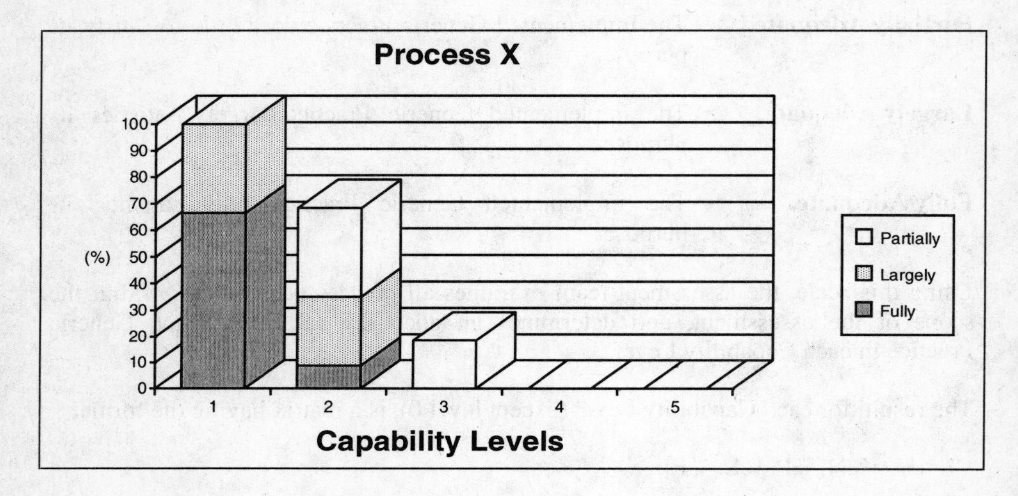

Fig. 7. A Process Capability Profile

It is important to note 2 major differences between SPICE and the famous SEI Capability Maturity Model (CMM) [13]. In SPICE:

1. a Process Capability measure is determined *for each process assessed* and not for the organizational unit

2. the result is not a single rating (e.g. the organization is at level 1 or 2, or whatever) but rather, *a profile* for each process assessed, indicating the achievements in each Capability Level. It is infact recognized that Generic Practices of higher levels may exist and be performed even when lower levels are not 100 % accomplished

7 Project status and future plans

At the time of writing (March 1995) the SPICE project has baselined some of its essential documents (PAG and BPG) and is about to complete the rest of the components. Plans have been made to have WG10 issue the complete set of products (standard and guidelines) as Working Draft (WD) in May 1995. The following 3 ISO

stages (Committee, Approval and Review) will take place between July 1995 and June 1997.
Meanwhile a world-wide trial has been undertaken, coordinated by the European Software Institute. Phase 1 of the trial will be used to test the Capability Model (BPG) and the Measurement Framework (PAG) and will produce a trial report for the end of May 1995.

8 Acknowledgements

This paper is based on the current results of the SPICE project contained basically in the set of documents listed in the references section. The author has attempted to summarize in a this paper the great amount of work accomplished by a large number of world experts composing the SPICE team. They are too many to mention and thank singularly but their contribution has been essential for the SPICE project and for the interesting parts of this paper. Thanks to them all. On the other hand, any mistake and/or uninteresting part of the paper is entirely my fault and I take all the responsibility.

9 References

1. ISO/IEC JTC1/SC7/WG7/SG1, *The Need and Requirements for a Software Process Assessment Standard*, Study Report N944R, Issue 2.0, 11 June 1992.
2. ISO/IEC JTC1/SC7/WG10 N002R, *Modus Operandi*, Ver. 1.01, June 1993.
3. ISO/IEC JTC1/SC7/WG10 N017R, *Requirements Specifications for a Software Process Assessment Standard*, Ver. 1.00, 3 June 1993.
4. ISO/IEC JTC1/SC7/WG10 N016R, *Product Specification for a Software Process Assessment Standard*, Ver. 1.0, 22 June 1993.
5. SPICE, *Project Overview*, Ver. 0.02 - Draft, 18 February 1994
6. SPICE, *Baseline Practices Guide*, Ver. 1.01 - 9 December 1994
7. SPICE, *Process Assessment Guide*, Ver. 1.01, 3 January 1995
8. SPICE, *Introductory Guide*, Ver. 0.06, November 1994
9. SPICE, *Process Improvement Guide*, Ver. 0.05, October 1994
10. SPICE, *Process Capability Determination Guide*, Ver. 0.04, November 1994
11. SPICE, *Assessment Instruments - Product Description*, Ver. 1.00, June 1994
12. SPICE, *Assessor Training and Qualification Guide*, Ver. 1.00, September 1994
13. Paulk, M.C., Curtis, B., Chrissis, M.B. an Weber, C.V. *Capability Maturity Model for Software Version 1.1*. Technical Report CMU/SEI-93-TR-24, Software Engineering Institute, Pittsburgh, Pennsylvania, USA, February 1993

Quality Estimation of Software Applications for Banking

M. Campanai[1], E. Ferretti[2], V. Valori[2]

[1]CESVIT High Tech Agency - Center for Software Quality
[2]CRF - Cassa di Risparmio di Firenze

Abstract. The paper will intend to summarize the results and the open issues derived from an initive of setting up a "management by metrics" program for the improvement of the software process, particularly the acquisition process of software applications for banking. The motivation to improve the software process resulted from a business need (i.e.: increase service profitability) and from external regulation. The analysis of an assessment of company's capabilities revealed the necessity of improvement. The adopted software improvement approach consists of an integrated collection of quality models, procedures, tools and training for increasing information system quality, improving the quality of each software components and controlling the development (internal or external) and the acquisition process. The results of these studies bear out the usefulness of starting with an assessment orientation for both process and product in an active approach to process management and improvement.

1. Introduction

The growth of software has focused on an "increased concern for quality" in software developers and users that appears and varies in proportion to the size and complexity of a project. Quality is becoming a must in the Information Technology area and a key issue when considering the services based on software. Many complex organizations need to manage software products and software related processes; the groups involved in the development, maintanance and servicing need strong confidence on the input /output they accept/deliver.

This paper deals with an approach to software quality improvement in software applications for banking started by Cassa di Risparmio di Firenze (CRF) in 1993 with the objective to improve quality in the larger sense, dealing with the whole business at a time, i.e.:

- to manage practices in software with a process model [3], [7], [8], in order to use and manage methods and techniques to analyze, design, develop, test, maintain, operate the software system.
- To ensure product quality characteristics for the business purpose [4].
- To ensure the adequacy of organizational aspects measuring key performance indicators.

The authors can be reached at the following address:
CESVIT, Fortezza da Basso, Viale Strozzi 1, Firenze 50129
e-mail: campanai@aguirre.ing.unifi.it

2.The Costs of Software

The company incurred high costs for in-house software personnel. In the past few year the software world has changed. Enhancements and new functions seem to be available from packaged software, meeting a broad set of needs and offering the opportunities of a high level of interperation, leading to systems free from proprietary conventions.

The company was dissatisfied with the in-house software, as all projects were late and over budget, and explored three possible strategies: (i) to improve in-house software building capabilities, (ii) the in market acquisition of packages, or (iii) outsource software development and/or maintanance buying applications and customizing them.

Build: The cost of building software currently runs from $ 400 per function point (FP) for small applications to more than $ 3,000 per FP for large applications. The average costs are too high and connected are the high risks of a continuous technological innovation. Many companies are exploring ways to improve their internal software performances, such as climbing the Software Engineering Institute maturity ladder [7], acquiring better methodologies, tools and technologies. Unfortunatly, very little empirical data exists regarding investments required to improve software performance, because for a long time companies lacked good software metrics.

Buy: A company that is only a marginal software-development performer has the choice to acquire commercial software packages rather than build its own applications. While results vary from vendor to vendor and package to package, the economics of buying commercial software seems to offer very significant economic advantages: Even mainframe packages such as the accounting systems have been dropping in cost, and some are available for less than $100 per FP.

Buy and Customize: The major drawback of buying software packages is that for large systems there is always a need to extensively modify a package after acquisition for guarantiing compliance and interoperability with the previous software environments (both software applications and data). The economics may become unfavorable; but this is usually the real choice the company has to make: to select a software package and to customize the application with outsourced resources.

2.1The Acceptable Software Quality Level

Although the principal issue is the quality's effect on cost, a general understanding of the quality's effect on income is required. The user's interpretation of "quality level" and the efficiency of IT based services are the basis for any price differential.

Moreover, as software maintenance represents a high percentage of costs for corrective, perfective and adaptive changes, the Quality Improvement Plan addresses the objective to assure fitness for use and assure technical quality, internal and external quality characteristics [4] with an Acceptable Software Quality Range [15], which means the capability to control a project within an optimal cost.

3.Software Process Improvement: the Motivation

The motivation to improve the process resulted from: 1) a business need (i.e.: increase service profitability), 2) external regulation, 3) the need to control the development of new projects in which it has been adopting new technologies such as visual programming, graphical interfaces and client-server architectures.

An appraisal of the software engineering practices from a software process management perspective identified the organization's strengths and weaknesses in the process; the informal process assessment also identified the values, beliefs, and unwritten rules that have shaped the organization. The assessment revealed a very low software process maturity level and a lack of an explicitely defined, measured, and controlled process. Generally, goals and time-related, measurable targets were not defined.

On the basis of this result, management committed the enaction of a quality improvement program aiming to improve the process, the product and the organizational aspects related to software. The program was defined on the basis of certain methods from literature such as the Quality Improvement Paradigm, Goal/Question/Metric Paradigm and the Experience Factory [10], [11] and the process assessment and quality improvement methods [5], [7], [8], [16].

3.1The Internal Assessment

The Process: The evaluation of some projects showed that, "essence difficulties" are inherent to software production [1] and that new technologies promoted as revolutionary help in the development, but they do not solve the inherent complexity of software. Moreover, for large systems development, software project management activities such as planning, staffing, organizing, measuring, visualization and controlling are extremely critical for the success of the project. The main pitfalls encountered were: inadequate system engineering during requirements analysis; inadequate tracing, tracking and management; improper sizing of target environment; selection of inappropriate methods and tool for the analysis and design phases of large systems; failure to provide metrics to track the progress of the project; lack of a continuous monitoring activity on system performance.

The Product: Notwithstanding the high number of packages and investments carried out, the problem with the deteriorating efficiency of certain processes used by agencies (real-time services for front office activities) has given way to a deep evaluation of the efficency of the information system, which has been confronted through monitoring the information system at the host level.

In late 1993, the QATP project has been started with the objectives of: gathering data about system load, census of applications, CICS monitoring and daily analysis of the load from the agencies, an impact of CICS in the operating system.

The results have demonstrated that the major criticalities were in the application environment, while on the CICS level, from a systemistic point of view, there would not have been any particular criticality.

3.2Supplier's Process Assessment

An informal assessment of suppliers' capability level was performed. The results highlighted, generally, a high level on the scale. Fig. 1 reports the results of an assessment using the Pressman method and tool [14], which considers 8 different process areas, (organization, methodologies and tools).

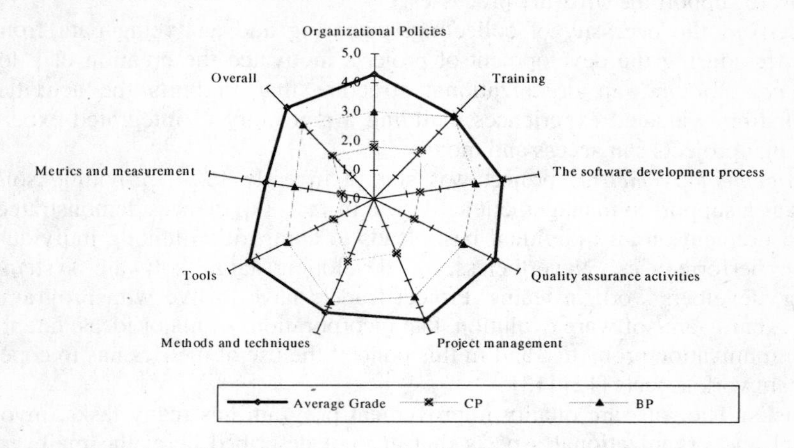

Fig. 1 - An example of a Supplier Process Assessment (8 areas + overall value)

The analysis of the assessment results of the suppliers and of some internal projects evidenced some interesting aspects: while the level of the suppliers was high, the level of the different projects, in which both suppliers and the company were involved was very low. The main problem was the "management of the project" in which more than one supplier was involved. In many cases, this came about because of an extreme faith in the supplier on the part of the internal personnel. Another aspect was that in large projects, the solution to a local problem often can create certain difficulties at the project level.

4.Process Improvement Program

The selection and successfull implementation of software improvements depends on many variables, such as the current process maturity, skill base, organization, and business issues and costs, risks and implementation speed.

The software process improvement was an integrated collection of software packages, procedures, tools, training and organization for increasing product quality and the control of the development/acquisition/installation/maintenance process in order to enhance the quality and the profitability of the service.

The defined Action Plan considered the following components to be essential: action-oriented description of tasks (for each action plan), responsability, resources, and schedule of checkpoints and milestones. An action plan typically has three parts: a strategic plan, one or more tactical plans, and one or more operational plans.

The first step was the definition of a taxonomy of software quality in terms of measurable software quality (process and product) models to define common and recognized quality characteristics and a minimum set of measurable indicators. The second step was to provide a model for software process enaction with defined and measurable activities to provide an organizational structure to support the development. The third was to define the quantitative basis for selecting methods and tools to support the software processes.

Moreover, the necessity of collecting, gathering and analyzing data from the experiences during the development of projects motivated the creation of a logical *Experience Factory,* an organizational structure that supports the activities by accumulating evaluated experiences, building a repository of integrated experience models that projects can access and modify.

A *management by metrics* project was started in early 1994, providing software metrics as a support to management activities; in fact, experiences demonstrated that the measurement on an individual basis leads to competition among individuals to improve performances. Nevertheless, in developing large software systems for banking, develpers work in teams. Project leaders have to live with programming turnover, hardware/software evolution, late incorporations of major ideas, latent bugs and communication problems, and in this context the use of metrics has to cope with other teamwork aspects [12],[13].

The Tasks. The software quality improvement program has many tasks, involving technical and organizational aspects that are not described here; the main aspects concerning software were collected in the following tasks:

1. Writing and distribution of the Software Quality Manual,
2. Definition of a software life cycle model
3. Definition and enaction of an Acquisition Process (software selection, customization, installation and maintenance)
4. Definition of Acceptance rules (software culture vs. contractual one) in accordance with the measurement apparatus
5. Application experiments.

4.1Software Quality Manual

The necessity to distribute a stable manual linked to the company needs and to the international standard was a constraint. The first step was the formalization of the actual software process. Many key process areas were formally undefined, but many people worked with quality criteria. The formalization in a quality manual of the quality procedures and the identification of technological constraints permitted the definition of a process and of the main procedures to guarantee a minimum level of quality. Standards and experiences from the literature have been evaluated and selected for the definition of the software process and for the definition of the quality model [3], [4], [5], [6], [9].

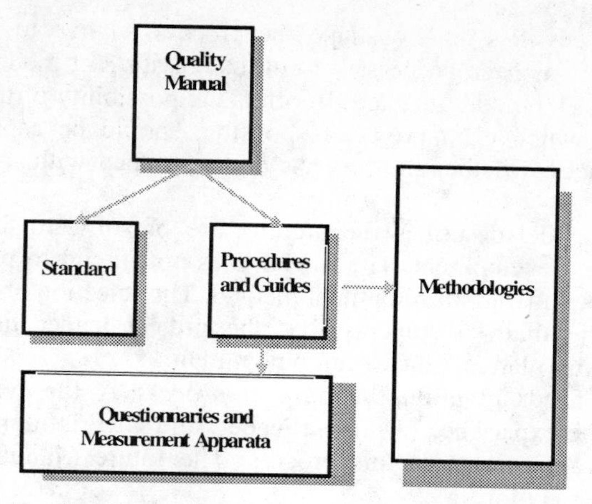

Fig. 2 - The Structure of the Software Quality Manual

The Software Quality Manual (see Fig 2 for the manual structure) has been intended to make life easier in project development. In the company there were many different projects in the organization mix, resources and technologies and tools. The main requirements of the manual were: i) the capability to model different processes that can be enacted for different projects (sometimes very different); ii) easy to use and to use; iii) capable to express a taxonomy of internal and external quality of software; iv) an approach based on the concept of "*project*".
Two different documents were adopted as main references .

- **Software process:** the model proposed in ISO CD 12207.2 [9] has been adopted for the capability of modelling the different types of projects in the Company (buy, make, reengineering,....) (see Fig 3).
- **Software product**: the ISO 9126 standard has been used as a reference for the selection of quality characteristics and sub characteristics [4], [6].

In the general framework of the Manual, some major aspects are stressed according to the priorities identified during assessment:

- The role of Operation Responsible (a sort of configuration management of operating software);
- The quality of data (to introduce a "data administrator");
- The classification of failure, errors and faults.

4.2The Life Cycle Model

A classification of projects highlighted that there were many different classes of projects with different objectives (from feasibility studies to the development of complex simulation models or distributed real-time applications). Variations in organizational policies and procedures, acquisition methods and strategies, project size and complexity, system requirements and development methods, among other things, influence how a system is acquired, developed, operated and maintained.

To meet these necessities and to shape the process relative to all the classes of project software, it has been neccessary to utilize an abstract model adaptable to the project demands. The model selected [9] offers the possibility of the approach based on "*project*", in which each project responsible should be capable of utilizing specific competences, whether the projects are developed with internal or external resources.

The ISO 12207 model describes the architecture of software life-cycle processes related to software development. The model does not intend to prescribe a specific life-cycle model or software development method. The selection of methods and tools are defined in the main documents that the project leader has to deliver and maintain: the quality plan and the development plan.

A special process needs attention, the *tailoring process*. At the moment, this process has been left to the experience of project leader, and CRF is monitoring projects to abstract the relevant aspects of this process. The future Manual will prescribe a detailed approach to tailoring.

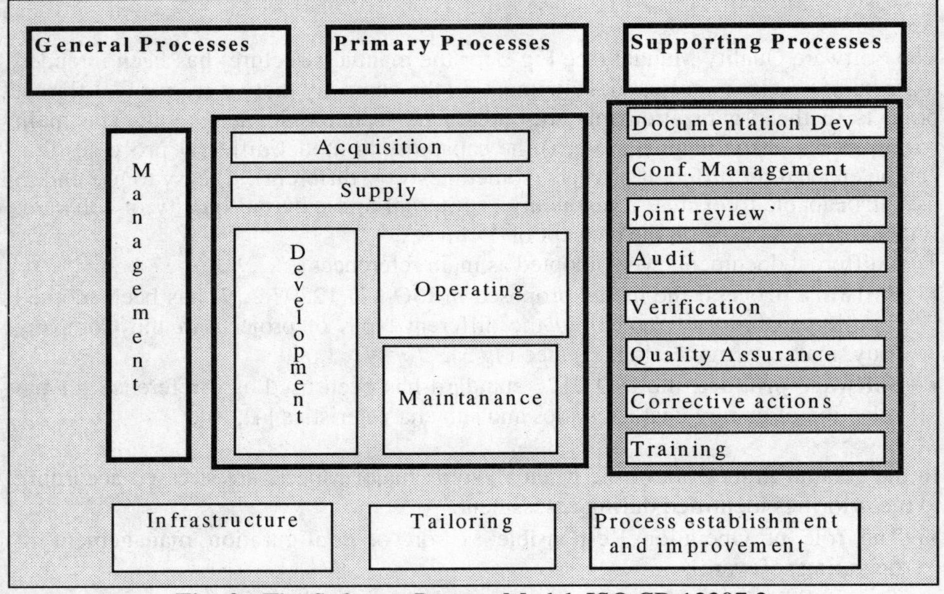

Fig. 3 - The Software Process Model, ISO CD 12207.2

The primary process has been defined as follows:

1. Acquisition Process; The acquisition process defines the activities and tasks related to the definition of the need to acquire a software product, the preparation and issuance of a request for proposal, selection of a supplier, and management of the acquisition process through the acceptance of the software product.
2. Supply Process; the supply process contains the activities and tasks of the supplier. The resources can be either internal or external, while the resource to manage and assure the quality of project are internal.

3. Development Process; The development process contains the activities and tasks of the developer: requirements analysis, design, coding, integration, integration testing and installation. The system testing is executed internally.
4. Operation Process; The Operation process contains tasks related to system operations and user support.
5. Maintenance Process; the maintenance process contains tasks related to modifications of the software system due to errors, deficiency and the need for improvement of funcional enhancements and external regulations.

4.3The Acquisition Process

The phase on which most of the attention has been focused on is about how to guarantee the quality of software systems during the acquisition. The process identified intends to make evident and usable to managers the dates relative to the products quality. In a particular manner, an acceptance model has been defined based on Quality profiles (Fig. 4). Two different questionarries have been defined: 1) with the objective of defining the profile of the quality long-awaited for the software product, and 2) with the objective to evaluate on a cheklist basis the measured quality profile of the software system.

The Acquisition process has been centered on three main activities: the definition of user and functional requirements, the definition of a software quality expected profile (Fig. 5) and the evalution of a measured quality profile (Fig. 6). The quality profiles have been defined for an easy to use approach of people from the organizational and management area with a particular attention to visualization of quality through pictures.

The definition of the expected Quality profile is based on a questionnaire with about 120 questions with a weighted impact on ISO9126 characterisctics. A study has been conducted to demonstrate a reasonable distributed impact of the questions on all characteristics. The expected results from the questionnaire are the profile, an evaluation of important constraints and the definition of a preliminary measurement plan, with an evaluation of the expected costs.

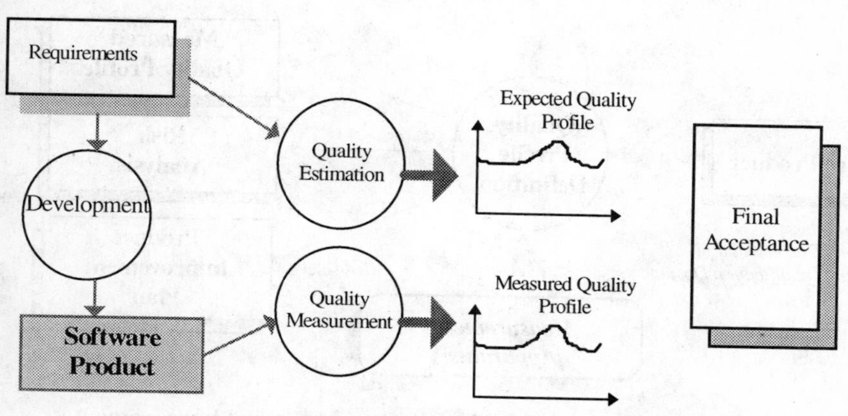

Fig. 4 - A Generic view of the Acceptance Process

The questionnaire is filled by users and analists, and its interpretation generates the exepcted quality profile. A tool for the automatic generation of the quality profile has been developed and distributed to each project leader.

The Acceptance process is based on software metrics and on the evalutaion of the measured quality profile. Three different areas can be investigated: the functional, the technical and the process oriented aspects. A questionnaire and a method for the selection of metrics (based on the Goal/Question/Metric approach) has been defined with about 400 questions. This is a more extensive one due to the contractual value of this assessment. The proof report show a quality measurement profile, that represents an interpretation of average values and a detailed evaluation of specific ties that represent measurments on a cost basis.

In this valuation phase based on cost, the two levels of valuation that identify the proofs that are to be carried out.

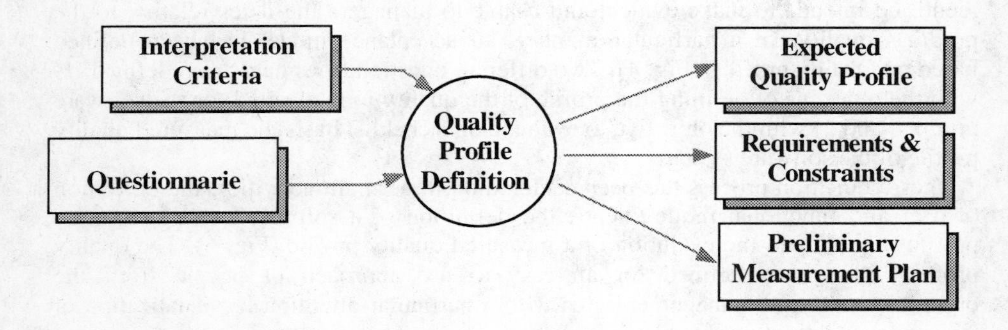

Fig. 5- The determination of expected quality profile

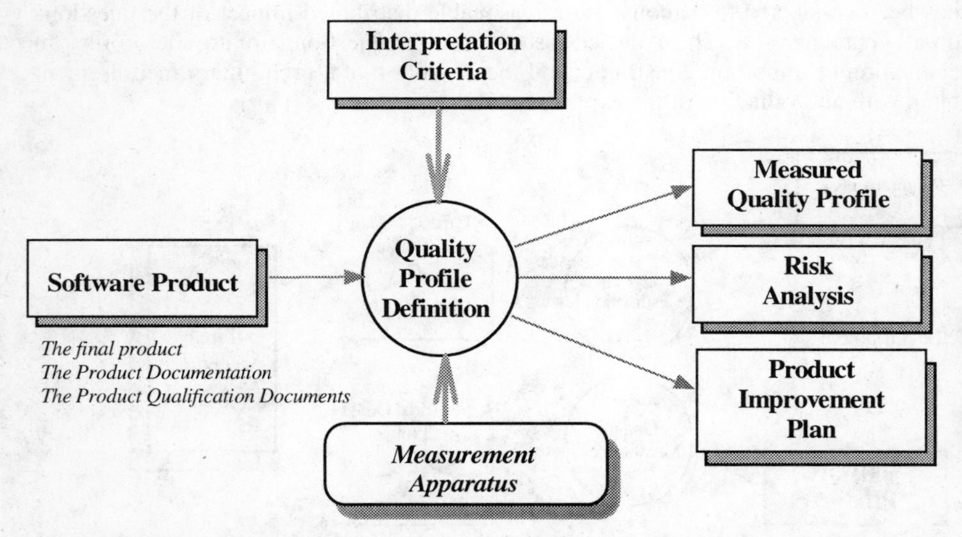

Fig. 6 - The Product Assessment for acceptance and Product Improvement

4.4The Measurement Apparatus

Measurement is the process of assigning a number to an entity to evaluate a specific attribute. The use of this number in a correct mathematical way is often very difficult: it is difficult to correlated complexity measure between different codes or between code and design documents.

The correlation of metrics with the same meaning and applicable to different products is a very important aspets of this approach. If the manager wants to track a project during its life cycle, it is a must to correlate different metrics with the same meaning. In this first approch to manage by metrics, we decided to use meaningful metrics, that is not to use true metrics [2]. The first objective of the measurement plan has to be the identification of metrics able to track the most relevant aspects of the project (both process and product oriented). This apparatus will be enhanced with the objective to be more validated and complete.

An objective of measurement is the low cost. The choice of some metrics can be very expensive; the selection of metrics takes into account this fact, and a selection of metrics according to the risk level and according to the available documents have been defined. Only applicable and meaningfull metrics are selected.

Each measurement plan investigates three different areas: the functional contents (adequacy), the technical (software engineering) and the process evidences (documentation and quality assurance). The method is a checklist based method; each question needs an inspection or measurement of the product. Questions can be Not Applicable. This permits the selection of metrics according to a defined method.

Some steps have been defined to get confidence on the metric plan:

1. Quality requirements definition (with the quality requirements priorities)
2. Attributes to be measured in terms of internal and external characteristics
3. Definition of acceptable rating levels
4. Definition of the measurement laboratory (tools and techniques) needed
5. Definition of the reporting requirements.

The measurement plan should be developed at the start of the project, and maintained during the development. Different technologies and different tools can change the objective of the measurement plan.

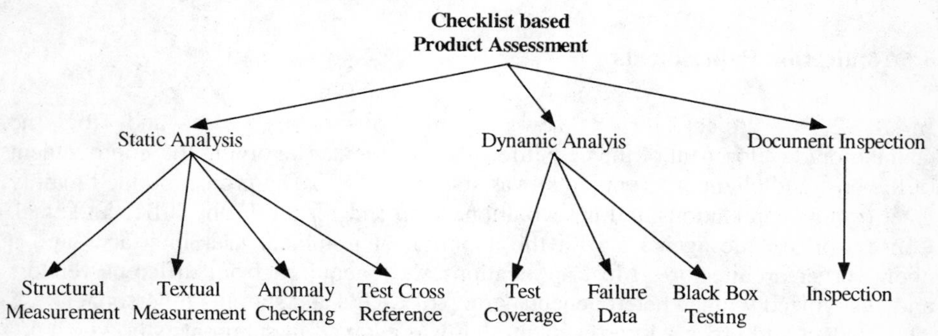

Fig. 7 - The Taxonomy of the Measurement Apparatus

A taxonomy of assessment techniques adopted for the evaluation of software product is reported in Fig. 7. All these techniques give some contribution to the evaluation of the quality model. The module used for the definition of measurement criteria is reported in Fig. 8.

Measurement	Functional aspects	***Y***
	Software Eng.	***Y***
	Quality System and Software process	***Y***
Classification of the product: Quality Profile, Risk Class	Functional class	Host Application A1 A2 Client-Server Applications Front Office Back Office .....
	Expected Quality Profile	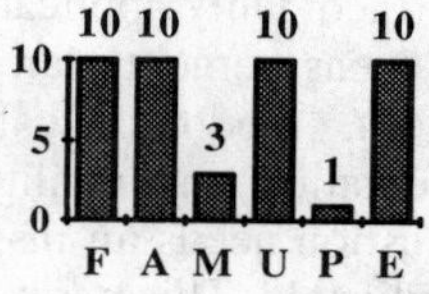 10 10 10 10 10 5 3 1 0 F A M U P E
	Risk Class	***B***
Measurement Accuracy	Functional aspects	***A Level***
	Software Engineering	***A Level***
	Quality System and Software process	***S Level***
Quality System		ISO9001 Certified

Fig. 8 - The Acquisition Process Standard Sheet

4.5Application Experiments

In 1992, a project for a new organization of agencies and for the acquisition/development of the new information systems (involving the improvement of the host and client environments) was started in CRF. The project defined mainly a set of host applications and an evolutionary project for the Front Office and Back Office work in the agencies with the adoption of graphical user interfaces and a client-server architecture. Most applications were acquired from different vendors and customized with an heterogeneous group of CRF's and suppliers' personnel.
The product and process improvement plan was adopted in subprojects in which 5/6 programmers and 5/6 testers were working. Different techniques for analysis and different languages were adopted: for some projects the languege was Cobol, in other projects a visual approach to software development was used.

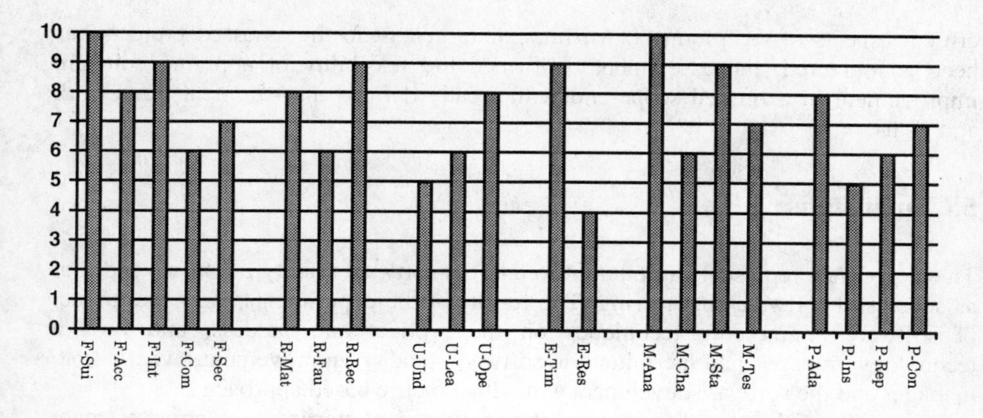

Fig. 9 - The expected quality profile (ISO9126 Sub-char level)

In Fig. 9 the *expected quality profile* for a subproject has been reported. In Fig. 10, an example of the functional evaluation sheet is reported. In this picture, there are some culomns with the following informations: the function (as defined in the requirements specification), the relevance of this function to the project, the presence of the function implementation, an evaluation of the criticality and a classification of the failures in terms of costs and relevance to functionality and reliability aspects (four levels of relevance M1-M4, G1-G4).

Function Name	Relevance	Coverage	Evaluation	Criticality	Functionality Reliability Modification
Census of physical person	3	present	2	3	G2 M2
Census of giuridic person	3	present	2	3	M2
Updating physical person	2	present	3	2	
Updating giuridic person	2	present	3	3	
Query physical person	2	present	3	1	
Query giuridic person	2	present	3	1	M2
Management of other and reserved information	3	present	3	3	
Insert of relationships among clients	3	present	2	2	G1 M2
Deleting relationships among clients	1	present	3	3	M2
Remove Inserted/Deleted relationships among ndg	2	present	2	2	M2
Query relationships among clients	3	present	2	2	M1
Search register	3	present	2	3	M1
Insert details ndg	3	present	2	2	M2
Delte details ndg	3	present	2	2	M2

Fig. 10 - The Functional Evaluation Sheet

The results of the evaluation: in these initial experiments, the evaluation process has been used in order to evaluate the quality of software systems and to assess critical aspects of the products. An immediate benefit to the assessed projects has been encountered; the assessment confirmed the feasibility of applying software improvement in a limited scope and with locally defined measurement framework and goals.

5.Conclusions

The experience represents an attempt to address software quality not as an end, but as successful software engineering. The results of the experiments show the impact of software engineering techniques on the product and process, and how a technology transfer had permitted a software quality improvement, where both manager and the software developers win with a metric based approach.

The results of these studies bear out the usefulness of starting with an assessment orientation in any active approach to process management and improvement and of using a metric based management approach to deliver software on time, within budget and with user's satisfaction.

Monitoring and measurements of professionals is used to provide feedback to help them improve their performances. Also, it is a sound basis for rewarding superior performance. As a negative point of view, people were "too ready" for the evaluation with the correct responses to the questions, and the assessors had great difficulty in discovering real weak points of the process and the product. Moreover, formal assessment is not useful for motivating improvements in organization. A bad result may represent a good reason to stop the improvement, while a continuous informal monitoring of the projects permits an interesting improvement of teams quality. At the moment, everyone accepted that a software crisis is not needed before an improvement initiative can take hold.

Future work: The experience lead to highlight future works such as: the adoption of ISO9126 model for the evaluation of the quality of data; the adoption of reliability models to assess the quality of suppliers monitoring the applications; the integration of methods for organizational aspects with methods for software development, the adoption of process modelling techniques for monitoring projects.

Acknowledgments: This paper reports the results of the work of all people involved in the quality improvement. A special thanks to A. Falzetti who committed the program, E. Grazzini, G. Orzati, M. Tozzi, A. Vicini, L. Dei, R. Trenti, A. Bandini, E. Greco, D. Palmieri, L. Rossato from Infogroup and the collaborative people from Olivetti and BankSiel.

References

1. F. B. Brooks, “No Silver Bullet: Essence and Accidents of Software Engineering”, IEEE Computer, April 1987, pp. 10-19.
2. N. Fenton, "Software Metrics, A Rigorous Approach”, Chapman & Hall, 1991.
3. ISO 9000-3, "Quality management and Quality Assurance Standars, Part 3: Guidelines for the Application of ISO 9001 to Development, Supply and Mainteinance of Software", 1991.

4. ISO 9126, "ISO9126 - Information Technology - Software Evaluation, Quality Characteristics and Guidelines for their Use", 1991.
5. ISO 9004-4, "ISO-IEC 9004-4 - Quality Management and Quality System Elements - part4 - Guidelines for Quality Improvement", 1993.
6. ISO/IEC WG6, Draft technical report of ISO/IEC JTC1/SC7/WG6, "Software Quality Evaluation Guide part 1,2,3,4,5,6,7,8" 1993.
7. M. C. Paulk, C. V. Weber, S. M. Garcia, M. Chrissis, M. Bush, "Capability Maturity Model for Software", Software Engineering Institute, Carnegie Mellon University, Pittsbourgh, Pennsylvania, February 1993.
8. Ami Consortium, "The AMI handbook", Ami consortium 1991.
9. ISO, "ISO CD 12207.2: Software Life Cycle Process", ISO, 1994.
10. V. Basili, D. Rombach, "The TAME project: Towards Improvement-Oriented Software Environment", IEEE Transactions on Software Engineering, Vol. 14. n. 6, pp. 758-773.
11. M. Oivo, V. Basili, "Representing Software Engineering Models: The TAME Goal Oriented Approach" IEEE Trans. on Software Engineering, Vol. 18, n. 10, October 1992, pp. 886-898.
12. T. Korson, V. Vaishnavi, "Managing Emerging Software Technologies: A Technology Transfer Framework", Communications of ACM, Vol. 35, N. 9, September 1992.
13. D. B. Simmons, "A Win-Win Metric Based Software Management Approach", IEEE Trans. on Engineering Management, Vol. 39, n. 1, February 1992.
14. R. S. Pressman, "A Manager's Guide to Software Engineering", Mc Graw Hill, 1993.
15. C. Hollocker, "Finding the Cost of Software Quality", IEEE Trans. on Engineerng Management, Vol. 33, n. 4, Nov. 1986, pp. 223 - 228.
16. T. Coletta, "The SPICE Project: An International Standard for Software Process Assessment, Improvement and Capabilities Determination", Proc. "Objective Quality '95", Florence, LNCS Springer Verlag, 1995. (This proceedings)

Validating Software Requirements Using Operational Models

Giorgio Bruno and Rakesh Agarwal

Dip. Automatica e Informatica, Politecnico di Torino
corso Duca degli Abruzzi 24, 10129 Torino, Italy
Email Bruno@polito.it, rakesh@polito.it

Abstract. It is well known that validating software requirements is an essential activity. Using operational models (i.e. models that are rigorous and mostly graphical and can be executed just as a very high-level programming language) can help the analysts get an insight into the system behavior and point out inconsistencies and missing requirements. This paper presents two modeling languages, Protob and Quid (the former covering functional and control issues, the latter addressing informational aspects) and illustrates their simulation and animation features, while emphasizing the architecture of models.

1 Introduction

The analysis of requirements is a critical activity. In fact, requirements which are not clear often lead to the development of inappropriate products, thus causing disputes to arise between the purchaser and the supplier. Therefore, if the analyst builds a model that formalizes the requirements, he or she can get an insight into the behavior of the system being developed and work out whether there are any possible inconsistencies or whether any information is missing, before the actual development takes place.

However, the study of a model yields only limited results if it is based solely on inspection, whereas if the model can be executed so that traces of the system's behavior are obtained, then a thorough analysis can be performed and the risk of delivering an unsatisfactory product is minimized.

Growing interest is being shown in operational models [1], i.e. models that can be executed using a suitable support environment. Most operational models are graphical and can be considered as high-level programs which are developed using high-level modeling languages.

An operational model can be modified and tuned until the behavioral traces it generates match those expected. In this way, the model is a reference point for the development of the system and, what is more, the purchaser feels confident that the system, being developed according to the model, will behave properly.

Operational models often allow timing constraints to be expressed and, consequently, a discrete-event simulation of the model can be performed. In this way, statistical estimates of the system's parameters can be collected in order to support decision making. When a formal proof would be too expensive, such statistics can confidently be used to determine some properties of the system.

Operational and evolutionary principles can be brought together to form a powerful software development paradigm. In fact, what characterizes software development with respect to the other disciplines, such as the design and manufacture of mechanical parts, is that the model and the final system have a common origin since they are both computer programs.

For this reason, the final system can be seen as the final step of an evolutionary transformation which enriches the initial abstract model with details and progressively turns it into the deliverable system.

This approach provides important benefits, such as

1. minimizing the risk of finding out that information is missing or inconsistent at the time the system is brought into operation;
2. maximizing the reuse of software modules (this is because their corresponding models are reused and reusing models is much easier than reusing programs);
3. improving productivity, because the final code can be generated from the model automatically.

This paper emphasizes the key role played by models. Software development consists, therefore, in building, testing and refining models within a seamless process that leads the analyst/developer from analysis to design and finally to the implementation of the system. Operational models can be executed and tested during each phase so that quality control is spread throughout the whole development cycle.

There are two key factors in the approach described in this paper.

The rigorous modeling language. The same language can be used at different levels (e.g. specification and design) and by different people (e.g. analysts, implementors and end-users) with unambiguous semantics. Since the language is operational, the model can always be validated.

Technology. Several advanced support tools are needed to put this approach into practice. The most important tools are graphical editors, code generators and simulation and application workbenches. Such tools enhance productivity as the complete application can be produced from its model automatically. Therefore, developers can avoid going deeply into the technical details of the underlying technology, just as programmers who use a third-generation programming language ignore all issues related to the operation of the processor and to the management of devices.

This paper presents two modeling languages, Protob and Quid (the former covering functional and control issues, the latter addressing informational aspects), and places great emphasis on the architecture of models. Software architecture is considered as a promising research area as pointed out by Perry and Wolf [2].

The paper also describes the simulation and animation of the models emphasizing the integration between the functional/behavioral part and the information one.

2 Protob

Protob is both a modeling and a development language for event-driven systems. It combines the most important features of high-level [3, 4] timed [5] Petri nets with those of extended dataflows [6, 7] and organizes them within an object-oriented framework.

A detailed description can be found in references [8, 9, 10], while the paper by Murata [11] is an excellent survey on Petri nets and the book edited by Jensen and Rozenberg [12] is a collection of recent papers on high-level Petri nets.

The application domain of Protob mainly concerns discrete-event concurrent systems, such as real-time embedded systems, telecommunications systems and manufacturing systems.

An application written in Protob is a collection of communicating objects, each object being an instance of a class.

Protob objects are also called *actors* to emphasize that they represent components which play an active role, as they can take decisions and react to external events autonomously.

A Protob class has a graphical part, the net: it is made up of *places*, depicted as circles, *transitions*, depicted as rectangles, and oriented *arcs*, which connect places to transitions and transitions to places.

Places contain units of information called *tokens*, which are mobile information packets. Tokens contain structured data (records) or references to objects.

A place can contain several tokens at a time; all the tokens contained in a given place are of the same type. Each place has three attributes: the place name, the place type and the number of tokens in the initial marking (such tokens are called initial tokens). The first two attributes are strings, while the third is an integer number which can be omitted if the place has no initial tokens. For the sake of expressivity, initial tokens are usually depicted in the illustrations as dots inside places.

Places are queues (not sets) of tokens, thus when a token is put into a place, it is added to the end of the queue. Tokens are ordered in places on the basis of their arrival times.

Transitions are the processing units of the model. They carry out token-driven computations.

If predicates, priorities and delays are ignored for the moment, a transition fires as soon as it is enabled (i.e. none of its input places is empty) and firing consists in removing one token from each of its input places and adding one token to each of its output places. The tokens taken from the input places are called input tokens, while those delivered to the output places are called output tokens. In Protob, there is no weight function associated with arcs, so, during firing, just one token is taken from an input place and just one token is delivered to an output place.

When a transition fires, it can execute an action. The action is a piece of C code which has visibility on the tokens acted on by the transition. The action can modify the contents of such tokens as well as invoke external subprograms.

Tokens are usually taken from places in FIFO order (i.e. the oldest token first) unless the transition has a predicate.

A simple example showing the interaction between a sender and a receiver is presented in figure 1.

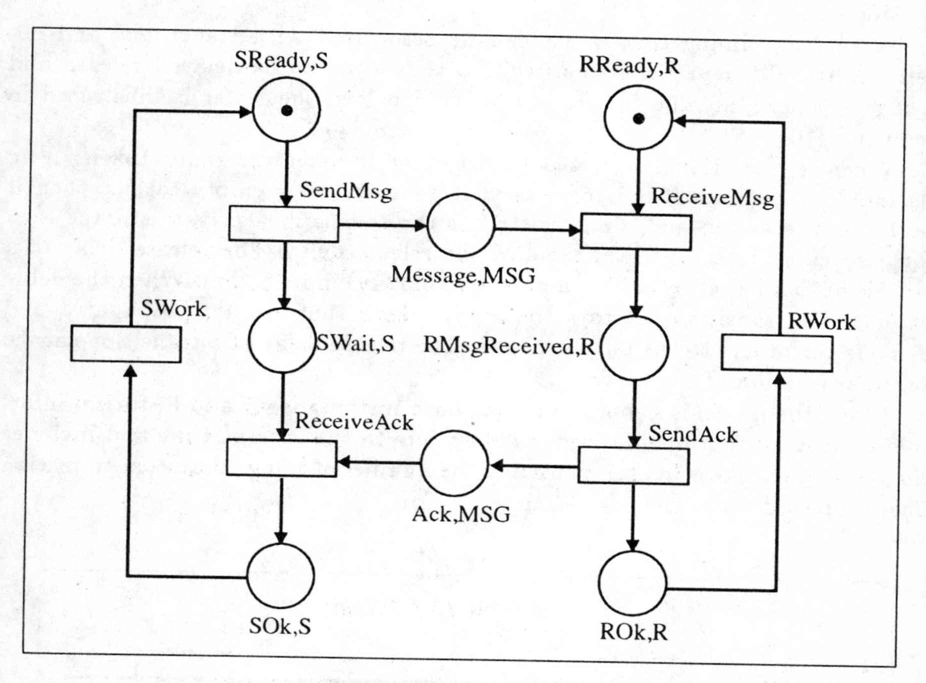

Fig. 1. The interaction between a sender and a receiver

1. Initially, there are two tokens in the net, one in place SReady and the other in place RReady, indicating that both the sender and the receiver are ready to start their activities.
2. In this situation, SendMsg is the only transition that can fire: it generates a new token (representing the message produced) and puts it into place Message, and moves the token from place SReady to place SWait. Now, the sender is waiting for the acknowledgement from the receiver.
3. Transition ReceiveMsg fires, thus consuming the token in place Message and moving the token from place RReady to place RMsgReceived.

4. After that, transition SendAck fires: it moves the token from place RMsgReceived to place ROk and puts a new token (representing the acknowledgement) into place Ack.
5. At this point both transitions ReceiveAck and RWork can fire. When ReceiveAck fires it moves the token from place SWait to place SOk and consumes the token in place Ack.
6. Transition SWork fires.

The introduction of timing constraints into the model enhances its descriptive power and facilitates a careful analysis of the performance of the system being considered.

In Protob, timing constraints can be associated with transitions and determine two different behaviors, which are referred to as delayed release and delayed firing. Only the former is described below, the latter is illustrated in reference [10].

When a transition is selected to fire, first it takes the input tokens from its input places and, if it is necessary, it generates new empty tokens, then it performs the action and, finally, it holds the tokens in a private storage area for a specified interval of time, called the release delay. The release delay of a transition can be set by calling a primitive (`delay`) in its action. When the delay expires, the transition destroys the input tokens that are not propagated and delivers the others to the output places. The release delay of a transition can be set in any action of the model.

When the model is being simulated, both instantaneous and historical information is collected. Instantaneous data refers to the current time and includes the number of tokens in each place and the number of firings that each transition has completed up to the current time.

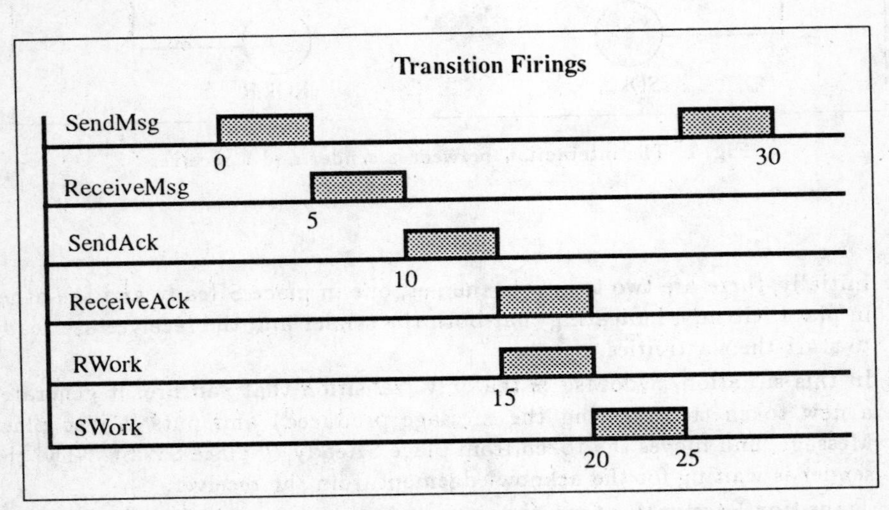

Fig. 2. The firing sequence in the interval [0..30] for the sender/receiver example

Historical data shows, over a given period of time, the firing sequences for the transitions on which attention is being focused. The diagram presented in figure 2 shows how the transitions of the sender/receiver model fire in the interval from 0 to 30 time units, assuming that the release delay of all transitions is 5 time units.

Models in Protob can be structured according to the principles of object orientation.

Since objects are based on nets, it is natural that they interact by sending and receiving tokens. Special places, called input ports or input places (of the object), and output ports or output places (of the object), are introduced so that an object is enabled to communicate with other objects.

Input places receive tokens from other objects. An input place is drawn as a double circle; it has a name and a type and contains a queue of tokens.

When an object has to send a token to other objects, it puts the token into an output place. An output place is drawn as a circle with a triangle inscribed. Output places do not hold tokens, because when a token is put into an output place, it is immediately delivered to the destination object(s).

The collection of all the input and output places of an object forms its interface.

Objects are graphically represented by a double square. Composition is graphically represented, too, because the model associated with a class can contain icons which represent objects belonging to other classes, as shown in figure 3.

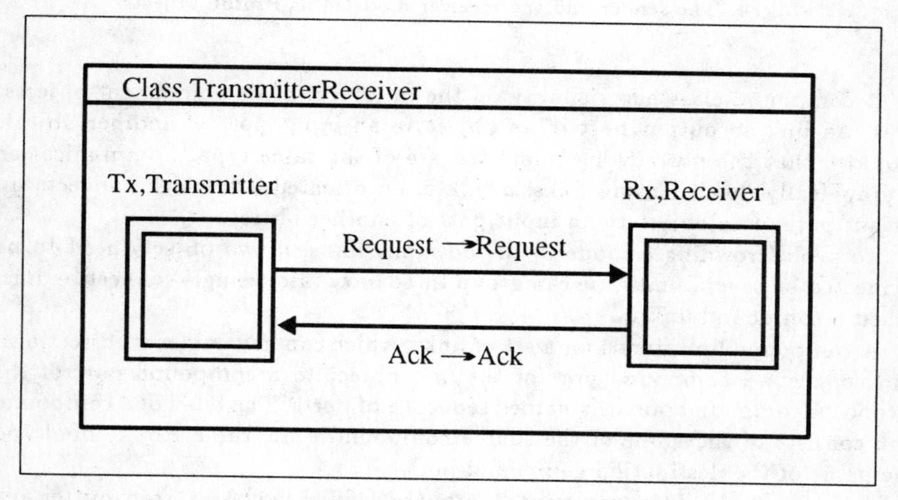

Fig. 3. An example of composition

For the time being, only fixed composition is considered, therefore a compound class/object has a number of components which cannot be changed.

An object has two identifiers: the first is the object name, the second is the name of the class to which the object belongs.

The classes, Transmitter and Receiver, of the objects, Tx and Rx, which appear in figure 3 are shown in figure 4.

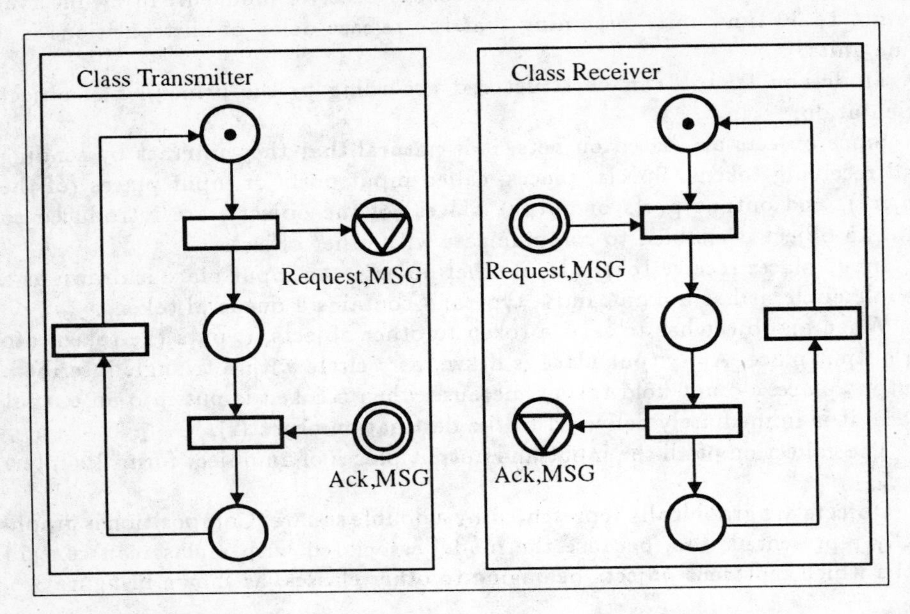

Fig. 4. The sender and the receiver modeled as Protob objects

A compound class has visibility on the interfaces of its component objects, so it can link an output port of an object to an input port of another object, provided that the ports to be connected are of the same type. Communication is graphically defined using links: a link is an oriented arc which connects an output port of an object to an input port of another object.

To avoid crowding a model with too many links, if two objects need to be connected by several links, we can group these links into a single connection line, called a compound link.

A compound link stands for a set of links, which can have different directions, and connects a compound port of the first object to a compound port of the second. A compound port is a named sequence of ports. The label of a compound link consists of the name of the source compound port, the arrow symbol and the name of the destination compound port.

As an example, if two compound ports are defined in classes Transmitter and Receiver (each compound port consisting of ports Request and Ack), a compound link can be drawn from object Tx to object Rx in figure 3 instead of two links.

Protob objects are easy to put together, in accordance with the metaphor of software chips [13]; in fact, an object does not know the other objects it will interact with and the interaction is only based on the tokens that it sends and receives through its interface, so it is the task of the compound class to set suitable links between its component objects.

If a model is large, it can be decomposed into portions (or submodels), called views. A view has a graphical representation given by a square with rounded corners. Views are exemplified in figure 7.

Other topics, such as using local variables and parameters, building client server models, extending inheritance to Protob nets, are not covered here for lack of space; the interested reader is referred to the textbook [10].

3 Quid

Quid is a language that has been designed to make Class-Relationship (CR) models [14] operational. CR is not generally provided with operational semantics, so there is no linguistic support that enables the analyst to work with objects and with associations between objects at a conceptual level. Consequently, CR models have to be translated into models based on lower-level languages, but this transformation causes a reduction in expressiveness, as discussed in [10].

Quid consists of a graphical part and a textual one: the former enables the analyst to build Quid models, such as the one shown in figure 5, while the latter enables the designer to generate objects and associations, to navigate among objects and to act on the objects reached during navigation.

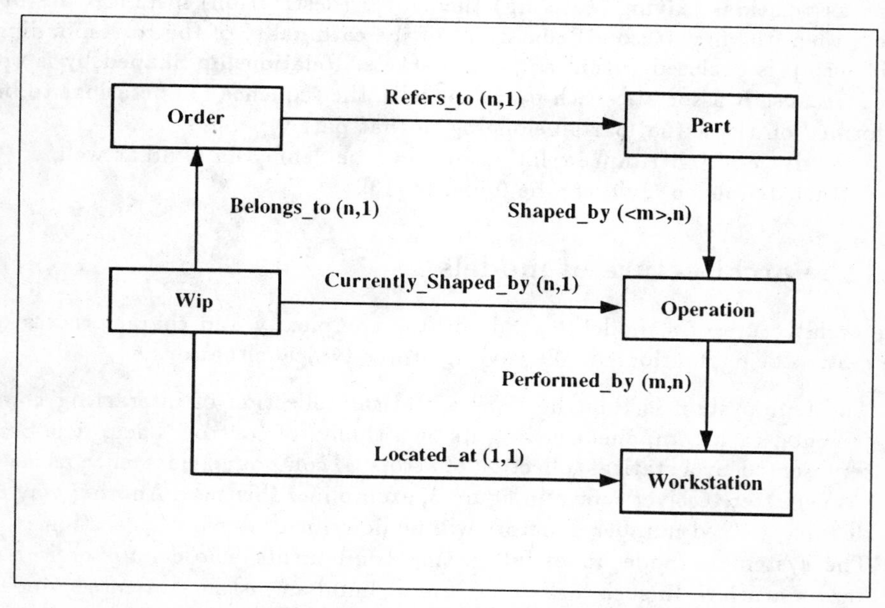

Fig. 5. The information model of a cell supervisor

Basically, a Quid model is a class hierarchy with binary relationships between classes. It is also called an *object model* because it represents the actual information structure (i.e. the actual object graph) consisting of interrelated objects. Navigation is the act of traversing the object graph following the paths that are specified using a navigational construct [15]. The objects that are reached during navigation can be acted on by calling the services of the corresponding classes.

The model presented in figure 5 represents the information structure of a cell supervisor. It shows that each order refers to a particular part type; each part type is shaped by a sequence of operations; each operation is performed by a set of workstations; each instance of class Wip, which models a part present in the cell (i.e. a component of the work in progress), is related to a particular order, is located at a particular workstation, and is currently being shaped by (or it has just been shaped by) a particular operation.

Relationships are drawn as oriented arcs for the arrow indicates the direction in which the name of the relationship must be read. For example, relationship Performed_by indicates that operations are carried out by workstations; Operation is the source of relationship Performed_by, while Workstation is its destination.

A relationship represents the associations that can exist between the objects of the source class and the objects of the destination class. Relationships can have cardinality constraints: one-to-one (1,1), one-to-many $(1,n)$, many-to-one $(n,1)$ or many-to-many (m,n).

Associations can be ordered and this constraint can be expressed as follows: if the associations leaving (entering) the source (destination) instances are ordered, then, the first (second) constraint of the cardinality of the corresponding relationship is enclosed within angular brackets. Relationship Shaped_by is ordered because it associates each part type with the sequence of operations to be performed on the actual parts belonging to that part type.

Recursive and inheritance relationships can be defined in Quid as well.

Further details on Quid can be found in [10].

4 The architecture of models

The architecture of a model depends on the complexity and characteristics of the system being considered. We envisage three typical situations.

1. The (sub)system is thought of as a statical collection of interacting components, each component having its own thread of control. Then, it is best represented by a statical collection of actors. A compound class, such as class TransmitterReceiver shown in figure 3, exemplifies this case. Another way of defining a fixed number of actors will be described later in this section.
2. The system is made up of interacting components whose number is not known a priori. In such a case, we cannot build an instance-oriented model, but we must adopt a class-oriented one. We use Quid in this case as well, but with a different interpretation: Quid entities are Protob classes and relationships are interpreted as compound links which connect pairs of compound ports. When Quid models are given such a meaning, they are called *actor models*.

3. An actor must act on a complex information structure and its action depends on external events. Then, the information structure is described by an *object model*, while the handling of external events as well as the actions on the information structure are performed by the actor, which includes the object model.

In general, an actor can contain other actors as well as actor models and object models. The system being considered is always represented by one actor, called the system actor; its class is called the system class.

4.1 Actors and object models

When an actor has to manage a complex information structure, it includes an object model, so the Protob net expresses the actor's behavior, while the object model represents a global information structure which is visible to all the transitions of the net.

The model in figure 6 shows the management and control of a manufacturing plant. A plant is governed by a plant supervisor and is made up of several cells. A cell is managed by a (cell) supervisor and consists of one warehouse, one cart and a variable number of workstations. Each device (warehouse, cart or workstation) is made up of two components, i.e. the controller and the machinery (which is controlled by the controller). Device controllers interact with the supervisor and, further, the cart controller communicates with both the workstation controller and the warehouse controller.

In figure 6, for the sake of simplicity, only the names of the objects are shown. Their corresponding classes are provided in a separate table.

The cell supervisor is an example of an actor that includes an object model, as shown in figure 7. The behavior of the supervisor is divided into 6 views. The graphical symbol of an object model is a single rectangle which contains two classes and one relationship. In this case, instances are passive components, while the instances of an actor model are active components.

An actor which includes an object model can take advantage of it in two ways:

1. Its transitions can include Quid statements in order to generate or cancel objects or associations as well as to navigate the underlying object graph.
2. Some tokens can be given the meaning of handles to objects in the object graph. This is done in the following way: if there are places in the Protob net whose type name is identical to the name of a class defined in the object model, then the tokens contained in such places are assumed to be handles to objects belonging to that class. In many cases the presence, in a given place, of a token which is a handle to an object indicates that the object is in a particular state. Therefore, the states of objects can effectively be shown using places without it being necessary to add state attributes to the corresponding classes.

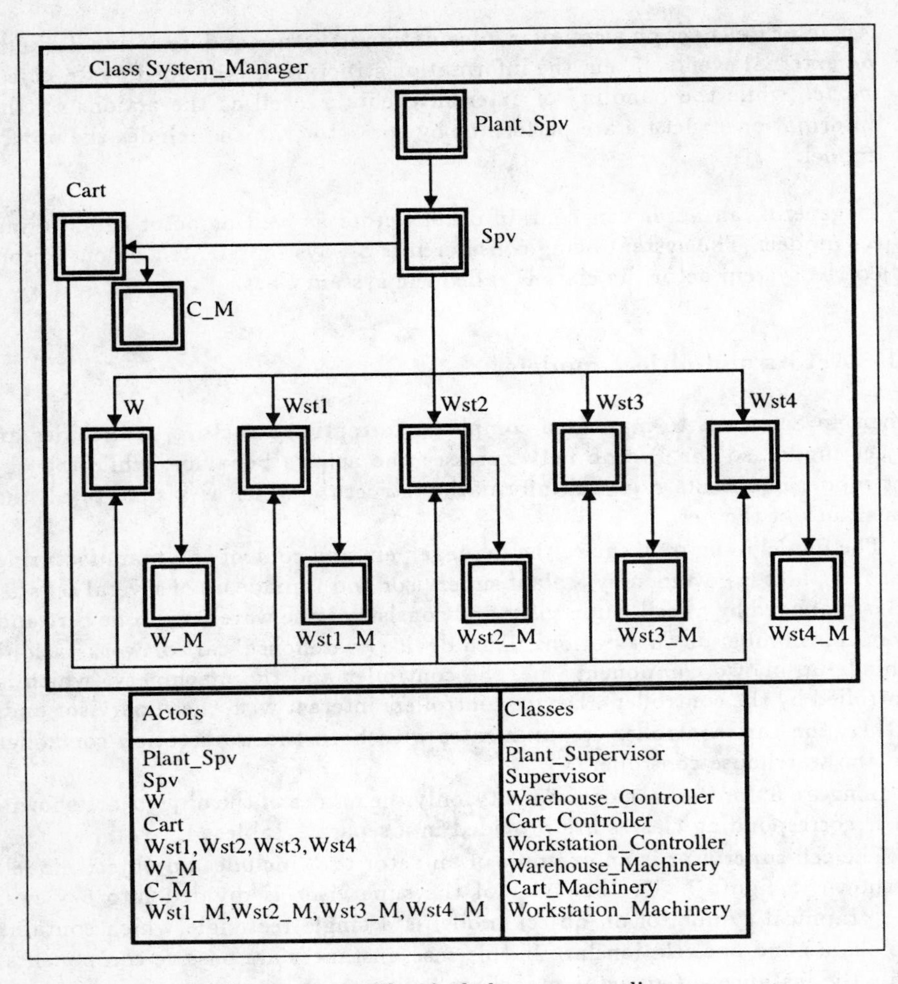

Actors	Classes
Plant_Spv	Plant_Supervisor
Spv	Supervisor
W	Warehouse_Controller
Cart	Cart_Controller
Wst1,Wst2,Wst3,Wst4	Workstation_Controller
W_M	Warehouse_Machinery
C_M	Cart_Machinery
Wst1_M,Wst2_M,Wst3_M,Wst4_M	Workstation_Machinery

Fig. 6. A class that includes an actor diagram

Object model Spv_Information_Model has been presented in figure 5.

A fragment of view Assign_Mission is shown in figure 8 and refers to the scheduling of missions of type 2.

A mission of type 2 consists in moving a part on completion from a workstation to another that is able to perform the next operation on the part. It can be started if there is a token in place Finished, which represents a workstation that has finished working on a part, *p*, and if place Idle contains a token representing an idle workstation that is able to perform the next operation on *p*. When the mission is started, commands are sent to the cart and to the destination workstation (through output places Cart_Cmd and Wst_Cmd). The token put into place M2 keeps the information to be used when the mission is completed.

The predicate and the action of transition Issue_Mission_2 will be commented on in the next section.

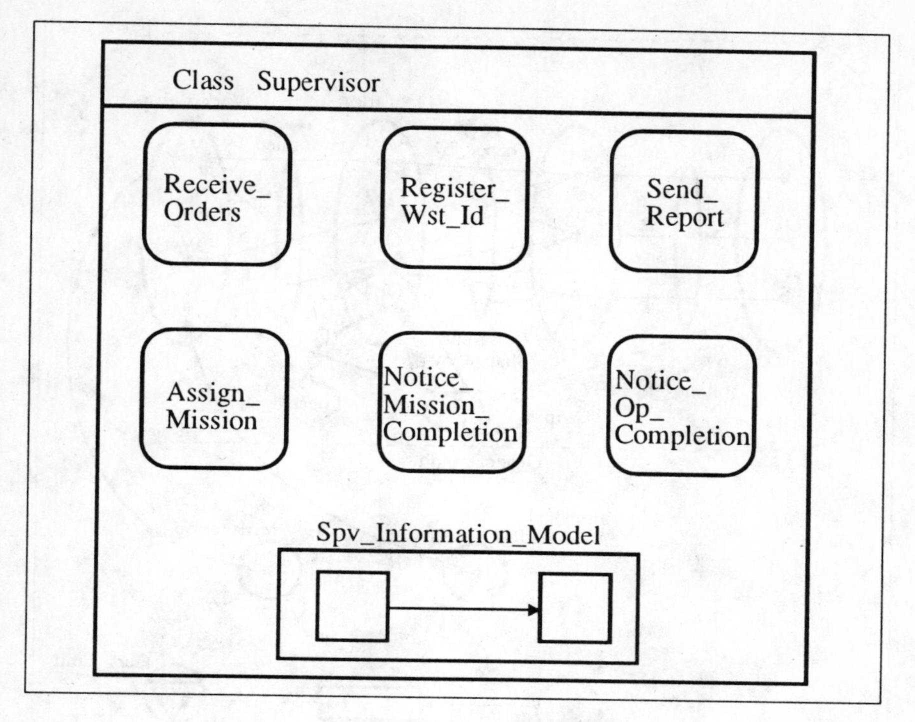

Fig. 7. The main view of class Supervisor

Places Idle and Finished contain tokens that are handles to objects of class Workstation and indicate the states of these objects.

5 Simulating and animating the model

Simulation is a well-known technique [16] that allows us to study how the model evolves over (simulated) time. Since the evolution of the model is of a discrete nature, attention is focused only on the instants at which something occurs (i.e. a transition ends its firing) and all the others are ignored.

Simulation can be run step by step or up to a certain time; using breakpoints, simulation can be suspended when a particular situation occurs.

During simulation several data can be collected and presented to the user as illustrated in figure 2.

It is also possible to show the contents of the tokens of interest at the current time.

The case in which tokens are handles to objects belonging to an object graph managed by Quid is more interesting, since it is possible to browse the object graph starting from such tokens. An example is given in figure 9.

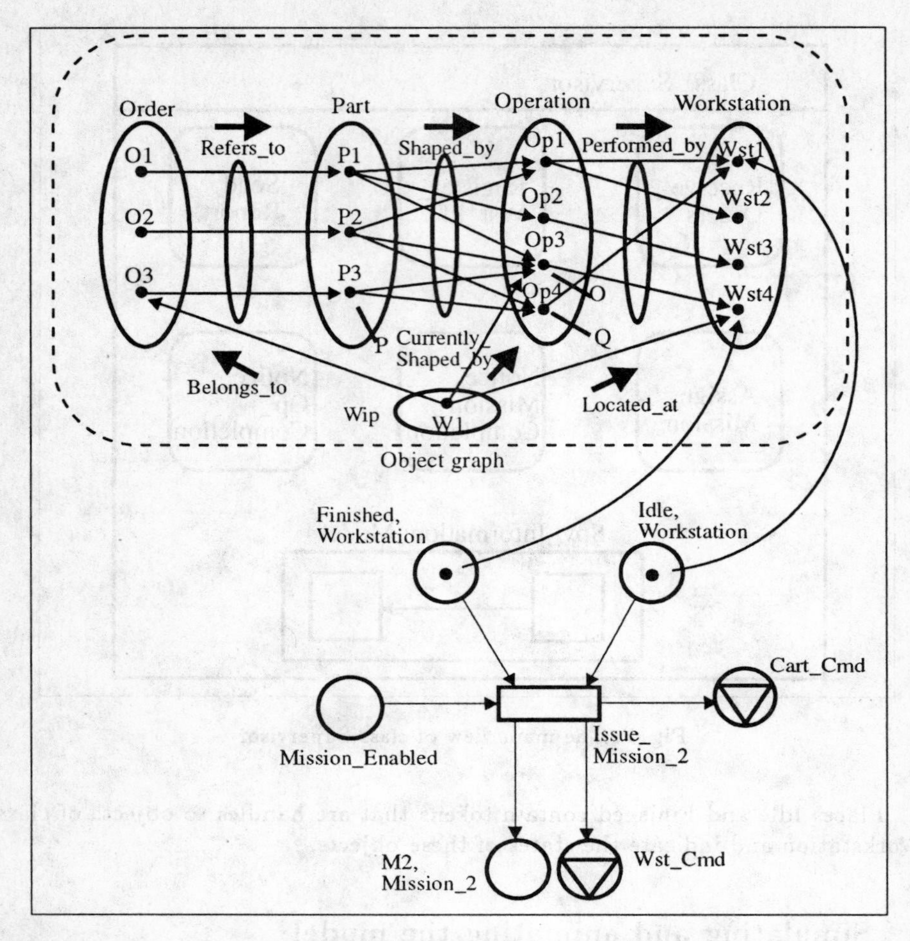

Fig. 8. A transition that acts on an object graph

At a certain instant, the object graph, whose model appears in figure 5, is assumed to be the one depicted in figure 8.

At that time, places Finished and Idle contain one token each: the token in place Finished refers to workstation Wst4 and shows that Wst4 has finished working on a part, whilst the token in place Idle refers to workstation Wst1 and thus indicates that Wst1 is idle. Now, if a token is put into place Mission_-Enabled, transition Issue_Mission_2 is allowed to fire only if Wst1 is able to perform the next operation on the part that is located at Wst4. This condition can be checked by associating a suitable Quid navigational construct with the transition. Details on this example can be found in [10].

However, we can check the condition directly by browsing the object graph interactively; this can be done (when simulation is suspended) by acting on the corresponding Quid model as shown in figure 9.

First, we have to select the object from which we want to start navigating the object graph. In this case, since we are observing tokens that are handles to objects, it is among such objects that we must select the one from which navigation will be started. In fact, we select Wst4.

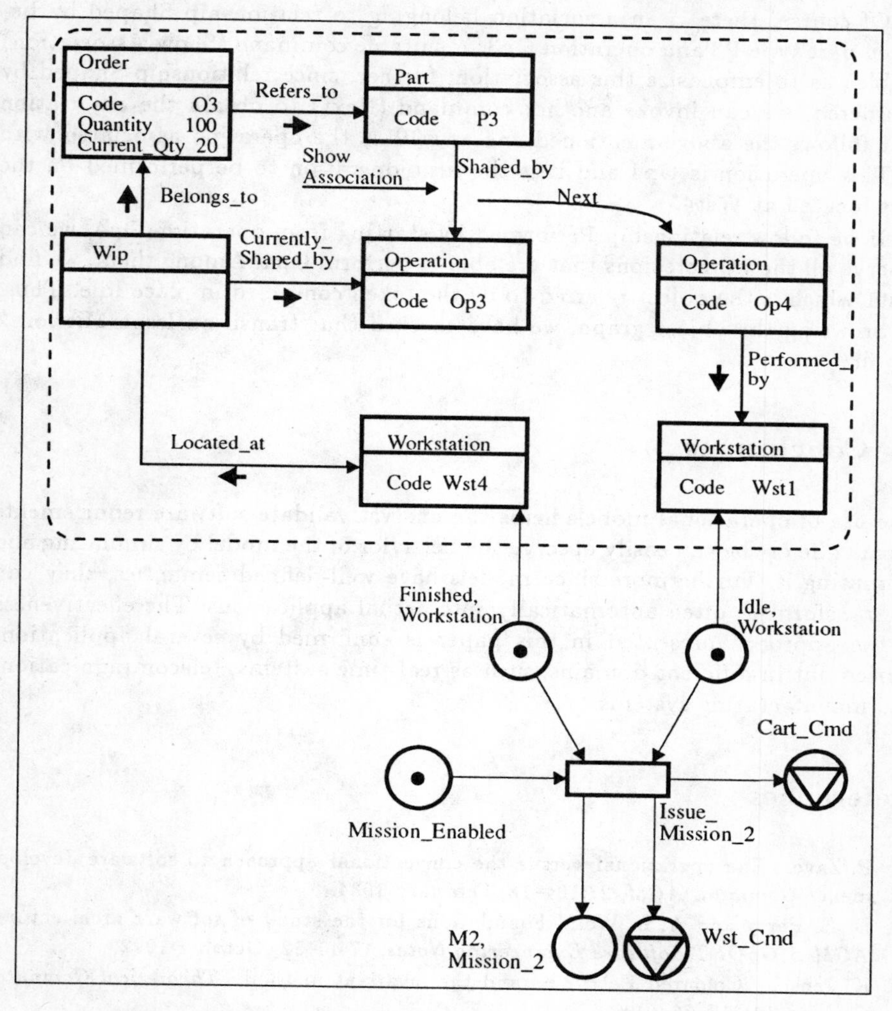

Fig. 9. Browsing an information structure during simulation

When we are focusing our attention on a particular object of a given class, we can examine the other objects that are associated with it by following the relationships in which its class is involved. In figure 9, an arrow near a relationship indicates that this relationship has been followed during navigation.

Therefore, if we start from Wst4 and follow relationship Located_at, we obtain the object of class Wip that represents the part located at Wst4. Then, following relationship Order, we reach order O3, which caused the production of that part, and, following relationship Currently_Shaped_by, we obtain operation Op3, which has been performed on that part. If we follow relationship Refers_to from order O3, we reach part type P3 to which the part located at Wst4 belongs.

Of course, there is an association belonging to relationship Shaped_by between part type P3 and operation Op3. A suitable command (Show_Association) enables us to emphasize this association; further, since relationship Shaped_by is ordered, we can invoke another command (Next) to obtain the association that follows the above-mentioned one as well as the operation associated with it. This operation is Op4 and it is the next operation to be performed on the part located at Wst4.

If we follow relationship Performed_by starting from operation Op4, we can observe all the workstations that are able to perform Op4. Among them, we find Wst1 which is the object referred to by the token contained in place Idle. Thus, by browsing the object graph, we have checked that transition Issue_Mission_2 can fire.

6 Conclusions

The use of operational models helps the analyst validate software requirements because he or she can easily observe the behavior of the model by simulating and animating it. Furthermore since models have well-defined semantics, they can be transformed, often automatically, into actual applications. The effectiveness of the approach presented in this paper is confirmed by several applications carried out in different domains, such as real-time systems, telecommunications and manufacturing systems.

References

1. P. Zave. The operational versus the conventional approach to software development. *Commun. ACM*, 27:104–18, February 1984.
2. D. E. Perry and A. L. Wolf. Foundations for the study of software architecture. *ACM SIGSOFT Software Engineering Notes*, 17:40–52, October 1992.
3. K. Jensen. Coloured Petri nets and the invariant method. *Theoretical Computer Science*, 14:317–36, 1981.
4. H. J. Genrich and K. Lautenbach. System modelling with high-level Petri nets. *Theoretical Computer Science*, 13:109–36, 1981.
5. C. Ramchandani. *Analysis of Asynchronous Concurrent Systems by timed Petri Nets.* PhD thesis, MIT, February 1974.
6. D. Hatley and I. Pirbhai. *Strategies for real-time system specification.* Dorset House, New York, 1987.
7. P. T. Ward and S. J. Mellor. *Structured development of real-time systems.* Yourdon Press, Englewood Cliffs, N.J., 1985.

8. G. Bruno and G. Marchetto. Process-translatable Petri nets for the rapid prototyping of process control systems. *IEEE Trans. Software Engineering*, 12:346–57, February 1986.
9. M. Baldassari and G. Bruno. PROTOB: an object oriented methodology for developing discrete event dynamic systems. *Computer Languages*, 16:39–63, January 1991.
10. G. Bruno. *Model-based software engineering.* Chapman & Hall, London, 1994.
11. T. Murata. Petri nets: properties, analysis and applications. *Proceedings of the IEEE*, 77:541–580, April 1989.
12. K. Jensen and G. Rozenberg, editors. *High-level Petri nets: theory and applications.* Springer-Verlag, Berlin, 1991.
13. B. J. Cox. *Object-oriented programming.* Addison-Wesley, Reading, MA, 1986.
14. J. Rumbaugh et al. *Object-oriented modeling and design.* Prentice-Hall, Englewood Cliffs, N.J., 1991.
15. G. Bruno, A. Grammatica, and G. Macario. Operational Entity-Relationship with Quid. In *Proceedings of the 5th International Conference on Software Engineering and its Applications, Toulouse, December 1992*, pages 433–42. EC2, Nanterre, France, 1992.
16. G. S. Fishman. *Concepts and methods in discrete event digital simulation.* John Wiley & Sons, New York, 1973.

Object-Oriented Design of the UPT Service: Evaluation of the Design Quality and Experience Learned

Giovanni Lofrumento   Sandro Pileri

Scuola Superiore G. Reiss Romoli S.p.A., L'Aquila, Italy

Abstract. In a competition market, quality has become the key issue for the survival of software producing companies. Software quality encompasses many topics. In this paper a design method quality is discussed and an overview of an object-oriented (OO) design method called Objects Design Method (ODM) is provided. ODM benefits of all quality factors of the OO paradigm and supports some quality concepts that systematically grant a better design quality. As an application of ODM, the design of the Universal Personal Telecommunication (UPT) service has been carried out and its relevant quality aspects and experience learned have been reported.

1. Introduction

Telecommunications (TLC) embrace a great deal of topics, e. g. switching systems, intelligent network, Broadband-ISDN, services, network management, transmission systems, etc., all controlled and managed through software by now. In the last few years, software has become the primary resource capable of quickly satisfying market needs in the TLC world. Both providers and telecommunications companies have established strategic goals to face the market competition. These goals are strictly linked to software, that is becoming the leading factor for developing and managing networks and services, and can be summarized as follows:

- widening of network services
- reduction in the period of introduction of new services
- shortening of services integration
- cost reduction in the development of new services
- cost reduction in the operation of both network and services
- better flexibility for network management
- improvement of services efficiency, reliability, and usability.

TLC software is characterised by largeness and complexity and requires thus good techniques for a proper management and design. Some of the characteristics required for TLC software design are:

- different abstraction levels
- modularity
- flexibility
- independence from hardware devices
- efficiency and reliability.

Software development is still far from the maturity gained in other engineering disciplines, but OO technology seems to have a good potential to solve many, even if not all, problems related to software. Most used OO analysis and design methods [1],

[2], [3], are enough suitable for the specification of small-medium size system; unfortunately, they provide little support for very large systems [4], especially for the systematic transition from OO software specification to OO software design, and often give vague indications about how to get a quality design. To improve software quality both quality processes and quality methods are considered necessary.

This paper focuses on quality methods by introducing an object-oriented (OO) design method called *Objects Design Model* (ODM) developed at SSGRR. This method allows a better quality design thanks to quality characteristics offered by the OO paradigm (modularity, flexibility, reuse, etc.). In addition, greater quality can be obtained as some more quality concepts are included in ODM as well. The major characteristics and quality concepts introduced by ODM can be summarised as follows:

- quality profiles required before the detailed design of architectural objects
- systematic transition from OO software specification to OO architectural design by using a reference model
- semi-formal textual and graphical notation for the object and dynamic models
- pseudocode for services specification automatically drawn out from detailed objects dynamics
- introduction of special design objects, called *port objects*, for the improvement of the system portability.

As an example of ODM application, a TLC service has been chosen and the quality of its design has been evaluated. The TLC service designed with ODM is the *Universal Personal Telecommunication* (UPT), which is not yet available in all countries.

The paper is structured as follows: the UPT service and the intelligent network (IN) are described; the main concepts of the OO paradigm are introduced; ODM is described considering only its architectural aspects due to limitation in paper size; the OO architectural design of the UPT service is provided considering the Italian IN structure; finally, some considerations about the design quality and the lesson learned are reported.

2. The UPT Service

Nowadays people want to widen their activities, either productive, commercial, cultural or recreative. As a consequence their needs of moving outside the boundaries that define their work or their culture are growing as well. The dimensions of their boundaries will probably grow to planetary proportion in the near future. TLC services will thus take on a main role and the reasons are twofold. The first is that the concept of *information productivity* is being widely accepted as it is the *business* coming out, directly or indirectly, from the use of TLC services. The second is the unwillingness to give up the same services when moving away from the usual environment.

These reasons have led TLC companies to make UPT service [15] available. The UPT proposal is that of a new service allowing both the access to TLC services and the personal mobility. Furthermore, each UPT user is allowed to access to a set of personalised services, to make and receive calls through different networks by using a unique, personal and network independent UPT number, and by using any terminal, either fixed, transferable, or mobile, without any indication about the geographic location. The implementation of such a service raises some problems, e. g. the alignment of different networks of different countries, or the harmonisation of distinct databases of different countries. Due to the many difficulties of the *complete* realisation of the UPT service, its introduction will be incremental. First the *restricted* UPT service will be implemented and then increased gradually up to the complete UPT. At least for its basic aspects, UPT is represented by two intelligent network services, the *calling card* and the *personal number*. The calling card service allows to make a call from any domestic place, public or private, charging a personal account; the personal number service allows a subscriber to switch calls from his usual place towards any place in the country where he is registered to. All the previous services require the intelligent network in order to be created, controlled, and managed.

2.1. The Intelligent Network

In the last few years many TLC companies have focused their attention on the development of *intelligent networks* (IN) both for the growing request for new services and for the cost reduction in building and maintaining services. The basic concept of IN is the centralisation of the services control logic and of the related databases into an *intelligent node* that provides services by interacting with other network elements. IN conceptual model is made of four planes at different abstraction levels that are *service plane*, *global functional plane*, *distributed functional plane*, and *physical plane*. IN distributed functional architecture [5], reported in Fig. 1, shows the IN functional entities and their interconnections.

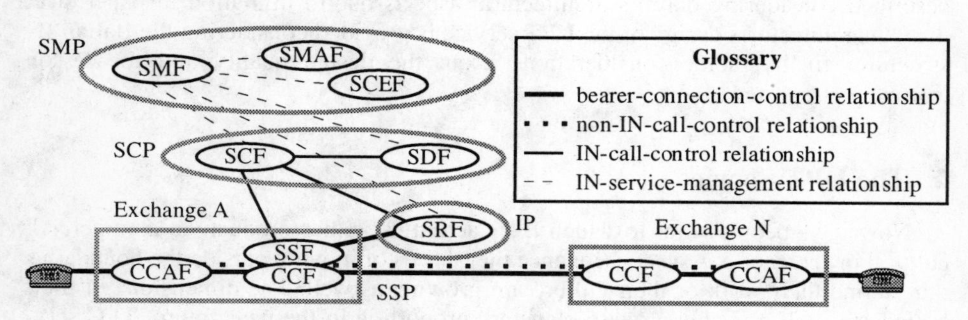

Fig. 1. Intelligent network distributed functional architecture

The *call-control agent function* (CCAF) and the *call-control function* (CCF) are allocated to the network exchanges and provide the basic functions for call handling. The *service-switching function* (SSF), allocated to an exchange called *service-switching point* (SSP), has the main task of detecting a request towards the IN. The SSF interacts with the *service-control function* (SCF), allocated to the intelligent node called *service-control point* (SCP), in order to control the service. The SCF communicates with both the *service-data function* (SDF) and the *service-*

management function (SMF) for the access to the service data and for the service management, respectively. The *specialised resource function* (SRF) provides advanced functions for the user such as voice recognition, voice announcements, etc. It is allocated to the *intelligent peripheral* (IP). The *service-creation environment function* (SCEF) provides facilities for service creation and testing and the *service-management access function* (SMAF) provides the management of the man-machine interface to the SMF. SMF, SMAF and SCEF may all be allocated to the *service management point* (SMP).

3. Basic Concepts of the Object-Oriented Paradigm

OO basic concepts and terminology are now explained before introducing the OO design model. The OO paradigm introduces a new way of modelling a software system that is seen as a set of interacting entities, called *objects*, rather than a set of functions. An object can be considered as an independent agent allowing a controlled access to its data through a set of services that represent its public interface. A software system can be, thus, modelled by a set of objects, without shared data, that have by definition a low coupling and a high cohesion. This guarantees a default quality level much higher than an equivalent functional model.

Basically, object characteristics and the interactions with other objects are reported in Fig. 2.

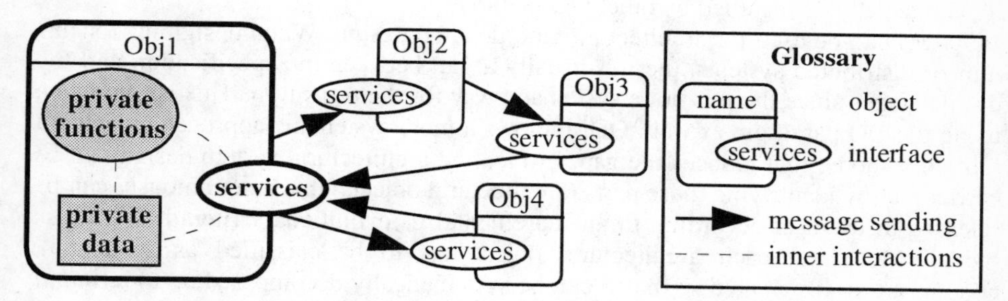

Fig. 2. Object characteristics and interactions among objects

An object is an entity that provides a mechanism for *information hiding*, that is it can have private data, called *attributes*, and function not available from the outside. What an object exports is a set of services to other objects.

Interaction among objects is made by message sending. A *message* is a request, carrying optional information, accomplished by an object towards another one. The object receiving the message executes the related service using the information carried by the message.

Usually an object is an *instance* of a *class*. A class represents a common template that defines characteristics and behaviours common to all its instances. Each instance, however, has its own copy of attributes defined in the class.

Classes can be defined as new or as a specialisation of existing classes using the *inheritance* property offered by OO programming languages. Finally, OO paradigm allows the use of *polymorphism*, that is the property that allows the different interpretation of the same message by different objects, each providing its own implementation. A complete description of all characteristics offered by the OO paradigm can be found in [6].

4. The Objects Design Model

The *Objects Design Model* (ODM) is a model of a methodology for OO software development that allows the architectural and detailed design of a software system for both static and dynamic concerns. To build the OO system design ODM requires an OO specification to be built through an object model and a dynamic model. The OO specification can be accomplished with any OO specification method, for example [1], [2], [3]. ODM is a natural continuation of the *Objects Specification Model* (OSM) [12] that can be used to build an OO software specification as well. OSM and ODM are integrated and are parts of a complete OO methodology still under definition at SSGRR. Concepts, formalisms, and notations used in OSM and ODM are not completely new; many of them have been extracted from both well-known software methodologies, not necessarily OO, and papers about the OO paradigm, e. g. [1], [2], [3], [8], [9], [10], [14], and have been integrated in an OO methodology in a coherent and consistent way.

ODM has not been defined as *another* OO design method, but as a pragmatic design method that also takes into account important aspects related to the design quality and usually disregarded by other OO methods.

ODM can be used for architectural and detailed design. When designing concurrent or distributed systems, there is usually a gap between the specification and the design stages since there is not a systematic way for the transition. This is true when using the OO paradigm as well. ODM allows a more systematic approach by using a reference model that reduces the gap. In fact, the architectural system design can be carried out by identifying those parts (architectural objects) that autonomously can be released to the user according to an incremental or evolutionary (iterative) process model. Moreover, each architectural object has to be classified as sequential, concurrent, or distributed so that it can be systematically decomposed up to *terminal objects* (objects no more decomposable).

Before its engineering into software, quality has to be specified [11]. The choices made during the system stage to find a solution, in fact, are strictly tied to the quality requested by the user. ODM requires the definition of a quality profile for each architectural object. The quality profile is very useful for a proper partitioning of each architectural object and for its detailed design. The method used to define quality, however, is not a part of ODM but either the ISO standard 9126 [13] or other methods can be chosen, for example that one reported in [11].

Another ODM characteristic is to isolate, when dealing with concurrent or distributed system, those parts that implement the communications among active objects. A new kind of design object, called *port object*, is introduced. Its aim is only the management of communication between active objects. Port objects allow the improvement of the system portability and maintainability.

Other ODM characteristics are its semi-formal graphical and textual notation for both the static and dynamic point of view and the possibility of getting the services specification, made by pseudocode, directly from the detailed dynamic model.

ODM is composed of a *Logical System Design* and a *Physical Nodes Architectural Design*. The former encompasses both a *software object model* and a *dynamic model*; the latter represents the hardware architecture made by one or more computers, where the software has to be allocated to, interconnected through local or wide area networks.

4.1. The Logical System Design

The object model of the Logical System Design (LSD) is based on the principle that a software system, at the architectural design level, can be viewed as an interaction of a set of software subsystems. In the ODM view, a software subsystem is considered as the minimal component that alone can be released to the user. A software subsystem can offer some functionalities and can interact with other subsystems. Each software subsystem is defined through a name and a set of services representing the public interface that it offers to other subsystems. In this view a software subsystem can be considered as an architectural object called *subsystem object*. Each subsystem object may be associated either to architectural objects or external objects. Each subsystem object at the implementation level will become an active object controlling all objects subordinated to it or that depend on it.

The system architectural design, then, is built through a set of subsystem objects that can be autonomously released to the user. Each subsystem object can be *sequential*, *concurrent*, or *distributed*. The number of subsystem objects that model a software system is a choice made during the architectural design phase. At first, the designer has not to keep in mind the inherent nature of the software system (i. e. sequential, concurrent, or distributed), but the system partitions that independently can be released to the user according to the *incremental* or *evolutionary* software process development. During the transition from software specification to software design, then, the objects contained in the specification object model are allocated to the subsystem objects. Once the object model has been understood and the architectural design has been accomplished with subsystem objects, there could not be any problem for the allocation of objects contained into the software object model. Subsystem objects detailed design can now proceed concurrently.

4.1.1. Sequential Subsystem Objects

A sequential subsystem object can be modelled by a set of lower level passive objects, either transient or persistent. A transient object does not need to have its attributes saved on a persistent storage; a persistent object, instead, needs to be saved each time one of its attributes is changed. Passive objects are related one another with the following relationships: *generalisation/specialisation*, *association*, *aggregation*, and *temporary association*. The notation used to represent such relationships is reported in Fig. 3 and is coherent with those introduced by OMT methodology [3].

The generalisation/specialisation relationship means that an object is obtained from another object inheriting all its characteristics and furthermore can have some more own characteristics. The association relationship means the two objects can communicate one another by message sending, without any kind of hierarchy between objects. The aggregation relationship is a stronger association between two objects, that is the aggregating object creates and controls the aggregated object. The temporary association relationship, not present in OMT, is related to another form of association between two objects that is used when two objects have to communicate temporary, that is only for the duration of a service of an object and not for its entire existence. For example, the last relationship is useful when persistent objects are temporary used by other objects (they are loaded from a persistent storage, operated upon by messages, saved to persistent storage and then discarded).

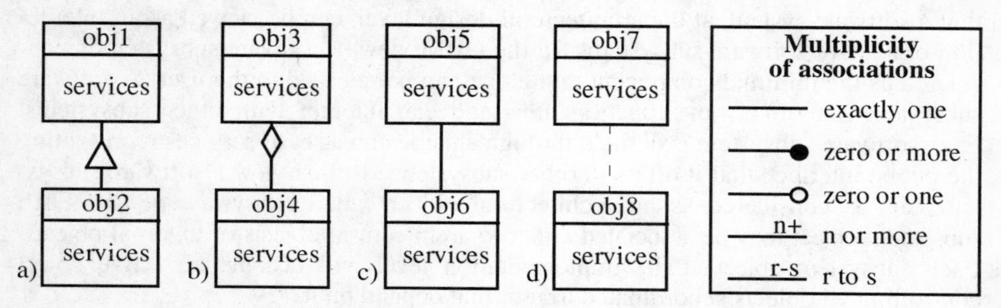

Fig. 3. Relationships among objects and multiplicity of associations.
a) generalisation/specialisation, b) aggregation, c) association, d) temporary association

Sequential subsystem objects are identified when the software system or one of its partitions is composed of either a set of passive objects that interact order to accomplish some services or a set of passive objects that are managed by a single active object. Each sequential subsystem object is represented by a single process when executed.

4.1.2. Concurrent Subsystem Objects

A concurrent subsystem object is modelled with at least two active objects. Each concurrent object can be *single thread* (STActiveObject), that is it has a single control flow, or *multiple thread* (MTActiveObject), that is it has two or more concurrent control flows. Each active object can be modelled by a set of lower level related passive objects as in the case of the sequential subsystem object. As the majority of OO programming languages are not concurrent, communication among active objects is accomplished by special objects, called *port objects*, that encapsulate the operating system mechanisms for the interprocess communication. A port object is represented by a hexagon near the object it is associated to.

Concurrent subsystem objects are identified when the software system or one of its partitions is made of independent components that have to be executed concurrently on the same computer in order to accomplish some services. Using a sequential programming language, each concurrent subsystem object is represented, when executed, by as many processes as the independent components it is composed of. Using concurrent languages or libraries and operating systems supporting concurrency (e. g. *Ada Task, threads libraries, Windows NT*), an independent component could be represented by a *lightweight process* that could be managed by either the operating system or the *run-time support*. The main difference for choosing an approach rather than another is related both to efficiency and reliability aspects and to an explicit design or not of the communication mechanisms among active objects.

4.1.3. Distributed Subsystem Objects

A distributed subsystem object is used to model a distributed system, that is a system allocated to at least two different nodes interconnected with a local or wide area network. The allocation unit is called *unit object*; at a lower level, then, a distributed subsystem object is modelled by at least two unit objects. Each unit object can be either concurrent or sequential. In the first case it can be modelled through one or more active objects and the modelling can proceed as in the case of a concurrent subsystem object. In the second case, the unit object can be modelled as in the case of a sequential subsystem object. Communication among distributed subsystem objects is accomplished by port objects that encapsulate all operating system and the network characteristics.

Distributed subsystem objects are identified when the software system or one of its partitions is composed of independent component that have to be executed on different computers, interconnected by networks, in order to accomplish some services. Each unit object can be represented in the same way as for concurrent objects when executed; the only difference is that in some cases the unit object could control the execution of the managed active objects.

4.1.4. External Objects

LSD also shows the external entities subsystem objects have to interact with. Such entities could be users, hardware devices, other existing software systems, and anything else outside the software system that interacts with it. In order to model external entities, *external objects* are introduced. Each external object is viewed as an object since it is described by a name and a set of services. Should an external object only send messages but not receive any message from subsystem objects, it could not show any service. External objects are associated to distributed or concurrent subsystem objects. At a more detailed level, external objects are associated to other objects (sequential subsystem objects, unit objects, STActive objects, and MTActive objects) by port objects.

External objects could be associated each other to show their connections.

4.1.5. Extensions of Active Objects

When designing a software system that has to be inserted into an existing framework or environment, some existing components could be modified in order to interact with the new ones. The design of the new component has to be accomplished considering the existing component and extending some of its functionalities. To model this situation generalisation/specialisation relationship has been assumed to be used according to its well-known definition and implementation when the components are passive. In the case of active components, instead, the previous relationship is not directly applicable; therefore an extended generalisation/specialisation relationship has been introduced. To say that an active object has to be updated according to the new system, it can be specialised with the active generalisation that graphically has the same shape as the conventional generalisation symbol, that is a triangle, but with an "A" inside. In the last description of the specialisation of an external object many details have been omitted for lack of space.

4.2. Structure of the Logical System Design

The structure of the LSD is shown in Fig. 4 through an object model. Objects with a single border are abstract, objects with a bold border are instantiable passive, and those with double border are active objects that have to be allocated to the nodes described in the Physical Nodes Architectural Design.

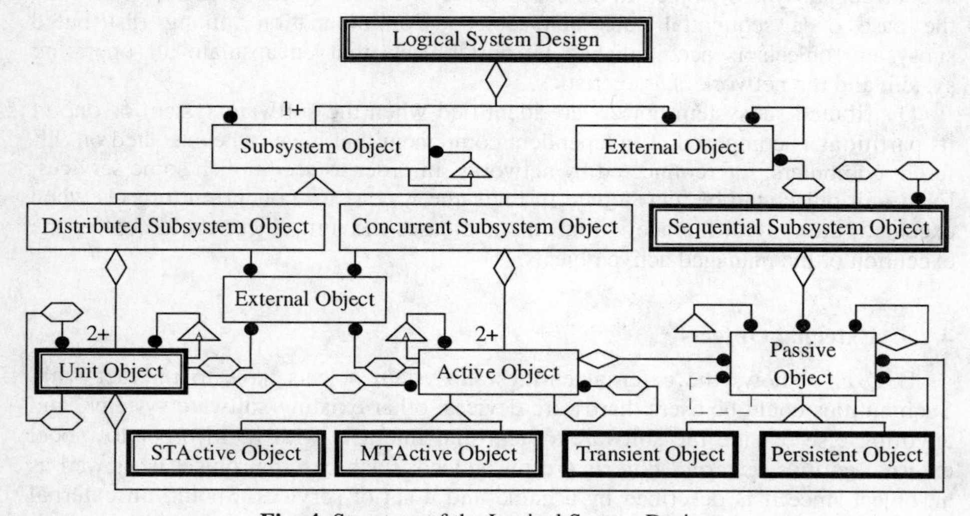

Fig. 4. Structure of the Logical System Design

The goal of the previous structure is to make a systematic passage from the specification object model to the design object model. According to the rules above specified the software system organisation is no more arbitrary but follows precise rule leading to a better architectural design.

Each object of the architectural design is drawn by a graphical symbol showing its name, its services, and the physical node where it is allocated to. Furthermore, each subsystem object is associated to the other subsystem objects, if any, it exchanges messages with. ODM is a semi-formal model as each object can be described by a textual script that has a precise and defined syntax. An example of graphical and textual notation of an object is shown in Fig. 5.

Each ODM object can be described with its own graphical and textual notation as the previous one. Sections of the textual notation included in squared bracket are not mandatory and depend on the object characteristics. Details on how to specify time constraints when dealing with real time systems can be found in [7].

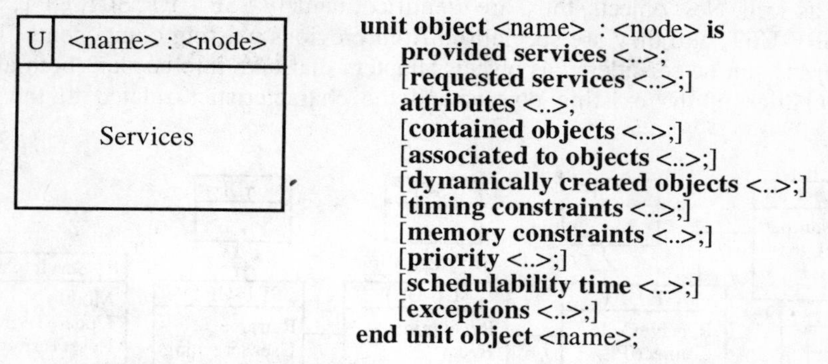

Fig. 5. Example of graphical and textual notation of the unit object

4.3. The Physical Nodes Architectural Design

The Physical Nodes Architectural Design (PNAD) shows the physical structure of the system in terms of nodes (computers and hardware devices) and links among them. PNAD provides the references where the subsystem objects have to be allocated to. Each hardware entity is represented by a node associated to another one; the association can have a multiplicity. Fig. 6 shows the PNAD structure through an object model. In the case of multiprocessor computers it is also possible to represent the processors of a node. In the software model, then, the specification that an active object has to be allocated to a particular processor would be possible.

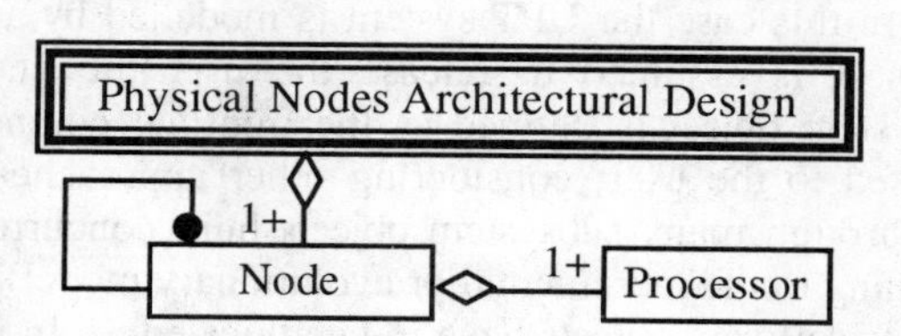

Fig. 6. Physical nodes architectural design structure

5. The Object-Oriented Design of the UPT Service

ODM characteristics are evaluated through the design of the UPT system. The UPT service is simple in its concept, but very complex when all its aspects are considered. The static and dynamic descriptions of the UPT architectural design is quite long and therefore only a brief summary, sometimes simplified, is given. First the UPT object model is built, as reported in Fig. 7, considering the macro-objects level.

The UPT system design is built into the IN framework and as such it has to be integrated with other IN software components, namely SSF, SCF, SDF, SRF, that control IN services. The existing components are modelled as objects with a name and a star that stands for their services. The reason is that since the existing components are not necessarily objects it is difficult to identify their precise services. However, the existing components have to be modified in order to manage the UPT service as well. New objects, thus, are identified, namely SSF-UPT, SCF-UPT, SDF-UPT, SRF-UPT, and they are specialised from previous existing components. These new objects can be considered as objects adapters that take into account both all the functionalities of the existing objects and the characteristics related to the UPT service.

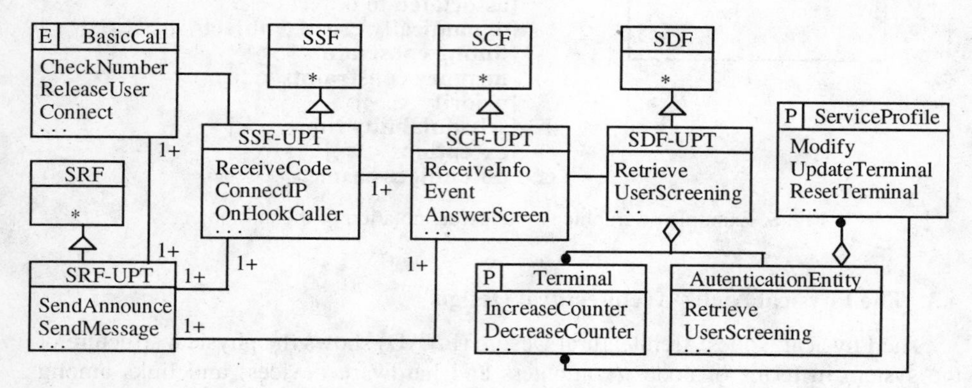

Fig. 7. UPT service macro-object model

The UPT system architectural design takes as input the previous object model and the dynamic model, not reported, and refers to the Italian IN structure. The UPT system architectural design is accomplished by both the LSD and the PNAD. LSD shows at the highest abstraction level the subsystem objects and the external objects of the UPT system; in this case the UPT system is modelled by a single distributed subsystem object as it is assumed to release the user the entire system at its completion (a subsystem object is defined as the minimal component that can be autonomously released to the user; considering other approaches the UPT system could be designed through many subsystem objects built concurrently and released incrementally according to the incremental or evolutionary model). PNAD shows the hardware components software objects have to be allocated to. In this case PNAD is homomorph to the structure of the Italian IN. UPT system LSD and PNAD are reported in Fig. 8.

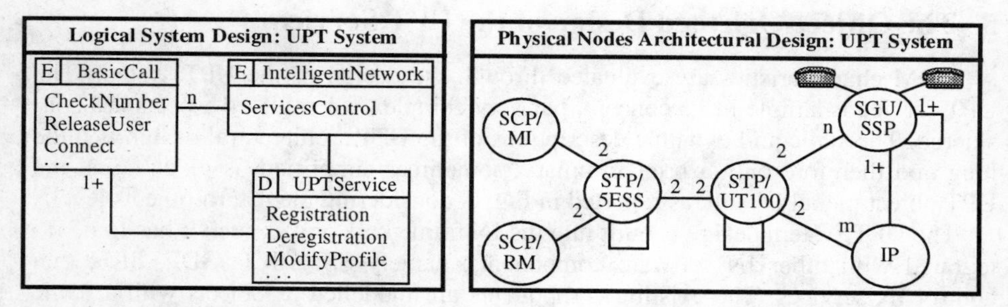

Fig. 8. Logical System Design and Physical Nodes Architectural Design of the UPT system

A distributed subsystem object is made of two or more unit objects, each one allocated to a physical node described in the PNAD. The UPTService subsystem object is mainly made of the unit objects reported in Fig. 9. There are also external objects (STP-UT100, STP-5ESS) and a reference object (BasicCall), that is an object described in another view, but interacting with some objects of the view. Unit objects are identified by the letter "U" in their left-top side. The unit object name is followed by a colon and the name of the physical node where it has to be allocated to.

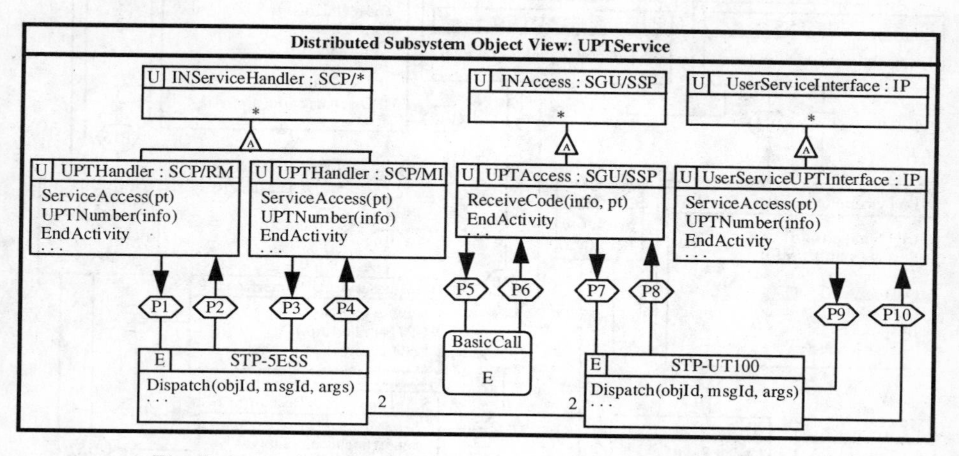

Fig. 9. Inner view of the distributed subsystem object UPTService

Communications among unit objects are accomplished by considering both the operating system and the network characteristics. Communications objects among unit objects have to be explicitly designed and are in any case managed by port objects drawn as hexagons. A port object could manage either unidirectional or bidirectional communication according to the choices made by the designer. To deal with existing software components that have to be augmented with new functionalities an extended generalisation/specialisation relationship for active objects represented by a new symbol shaped by a triangle with an inner "A" is introduced. The reason is that the implementation of the generalisation/specialisation relationship among passive objects is easily made by the inheritance mechanism of an OO programming language. Such a mechanism is not directly applicable when the objects are active since some more explicit, often not systematic, actions have to be done by the designer (such considerations are true assuming a sequential OO programming languages for the system implementation).

Each unit object reported in Fig. 9 is modelled, according to the rules of the reference model shown in Fig. 4, by more specific inner objects in order to have a more detailed model. The UPT system architectural design, thus, can be obtained by merging all views related to each unit object. In Fig. 10 a first level architectural design is reported. It is composed of all unit objects as well as the main concurrent and passive objects.

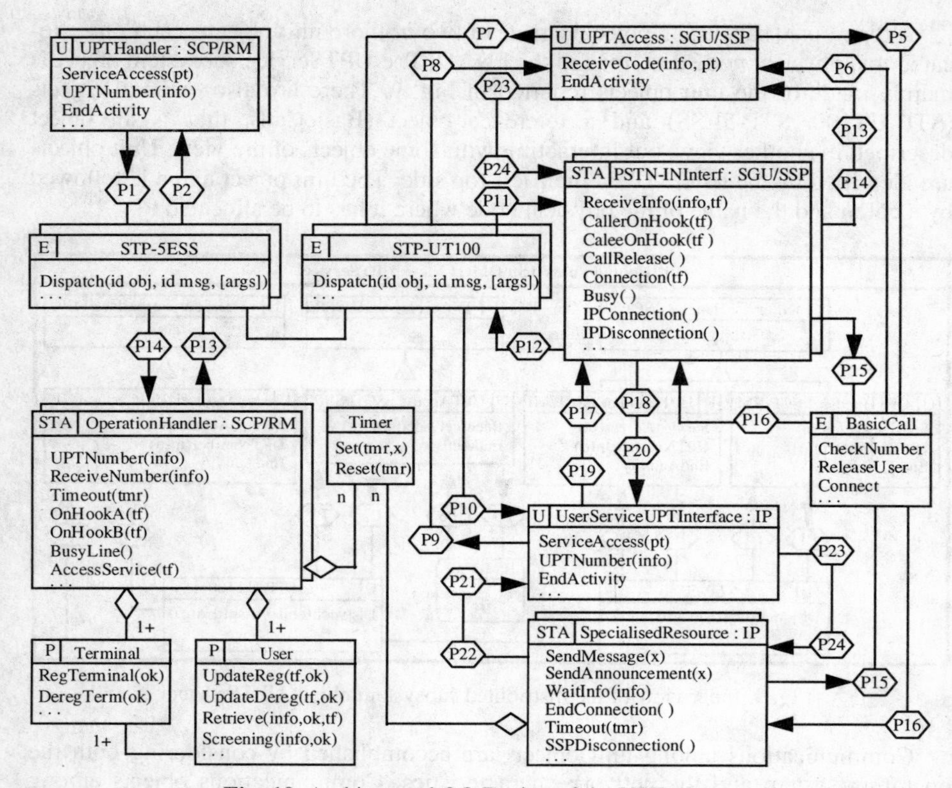

Fig. 10. Architectural OO Design of the UPT system

Detailed design continues again according to the rules of the reference model shown in Fig. 4 until terminal objects are identified and related one another.

The UPT system dynamic model is modelled by both *message sequence charts* (MSC) and *Objects Behaviour Model* (OBM). MSC shows inter-objects dynamics and OBM is used to model the inner object dynamics. OBM is built at two abstraction levels. The former, called *Object States Diagram* (OSD) partitions an object behaviour through a set of states according to the rules of *finite state machines*. The latter, called *State Actions Diagram* (SAD), shows the detailed actions accomplished within each OSD state. Examples of both an object states diagram and a state actions diagram of the object *SpecialisedResource* are reported in Fig. 11 and Fig. 12, respectively.

6. Evaluation of the UPT Design Quality and Experience Learned

In this section some evaluations and experience learned about the use of ODM are reported. A specification of the UPT service has been previously carried out with an object model and a dynamic model according to the rules of OSM. As far as the transition from the specification to the design stage is concerned, ODM introduces a structure for a systematic passage that both reduces the designer subjectivity and improves the design. Moreover, ODM is, in some sense, tied to a software process model. In fact, it requires some choices about the pieces of software that have to be released to the user. In any case, ODM provides a support for every software process model, e. g. waterfall, incremental, evolutionary, or spiral model. The choice of a single subsystem object at the architectural level means that the global system has to be released to the user and this implies something like the waterfall model. An architectural design made by many subsystem objects means an implicit choice of either the incremental or evolutionary model as the software will be released gradually to the user. The shift from a process model towards another one could be established by risks evaluation made during the software project management.

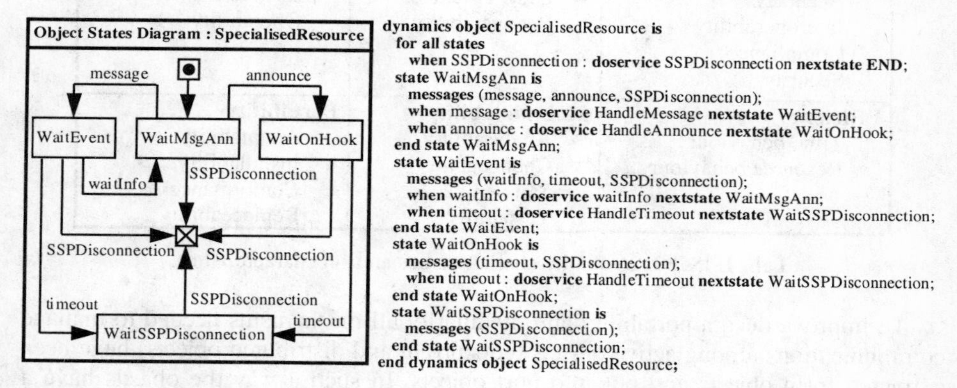

Fig. 11. Graphical and textual description of SpecialisedResource object dynamics

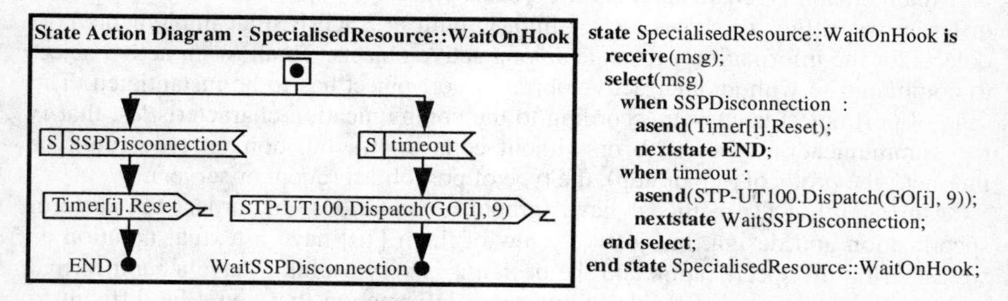

Fig. 12. Graphical and textual description of a SpecialisedResource state

Referring to quality characteristics, they have to be explicitly specified before their engineering into the software system. The association of a quality profile to subsystem objects has proved to be very useful as it has driven the choices for detailed design. In fact, decisions for detailed design and implementation are strictly tied to the specified system quality. For each subsystem object the quality profile has been determined according to the ISO standard 9126 that suggests six quality characteristics, namely functionality, reliability, usability, efficiency, maintainability, and portability, each subdivided into some more quality subcharacteristics. A subsystem object quality profile has been determined giving a value, in a range from zero to five, to each quality characteristics, where zero means that the quality characteristic has not to be considered at all and five means that the maximum effort has to be accomplished in order to engineer the characteristic into the software. When necessary, quality subcharacteristics have been considered as well. Some quality characteristics, however, could be conflicting one another so a matrix with conflicting and complementary relationships among quality characteristics, like that one reported in [11], has been considered in order to adjust such characteristics. ISO 9126 quality characteristics and subcharacteristics are reported in Tab. 1.

Functionality	**Reliability**	**Usability**
Suitability Accuracy Interoperability Compliance Security	Maturity Fault Tolerance Recoverability	Understandability Learnability Operability
Efficiency	**Maintainability**	**Portability**
Time behaviour Resource behaviour	Analysability Changeability Stability Testability	Adaptability Installability Conformance Replaceability

Tab. 1. ISO 9126 quality characteristics and subcharacteristics

To improve design portability and modularity all mechanisms needed to manage communications among active objects (concurrent and distributed objects) have been extracted from objects and put into port objects. In such a way the objects have a better cohesion as well and the quality of the system is improved as far as portability and maintainability characteristics are concerned. Port objects can be designed in different ways. Fig. 13 shows a class library, built at SSGRR, that implements port objects for the information exchange among active objects. To allow an active object to communicate with another active object, a port object has to be instantiated. The kind of port object is chosen according to the communication characteristics, that is the communication type (with or without connection), the domain type (unix or internet), the protocol (tcp or udp), the type of port object (client or server).

Almost all OO methods have graphical notations in order to perform specification and design, but just very few of them [10] have a textual notation as well. During the specification and the design it is very useful to have a semi-formal textual notation as it is possible to add more information that could be difficult to represent with the graphical notation. Such information could be related to the specification of time constraints, to the description of the states of an object, or to the description by a formalised pseudo-code for the behaviour of objects services.

By using ODM a support to all the previous items has been observed with an obvious benefit of design quality. ODM will be further experimented for designing real problems and will be properly tuned up to get a design method that allows the achievement of better design quality in a systematic way.

In conclusion, during the UPT system design made by using ODM the following characteristics have been observed:

- easy integration with many OO specification methods
- support for many software process models

- many different abstraction levels in order to manage complexity
- systematic transition from OO specification to OO design
- better detailed design driven by quality profiles
- high modularity providing a design with much more visibility and flexibility
- introduction of special objects for concurrent and distributed communications to improve the system portability and maintainability
- pseudocode for services specification automatically extract from detailed dynamic model
- design structure resilient to changes.

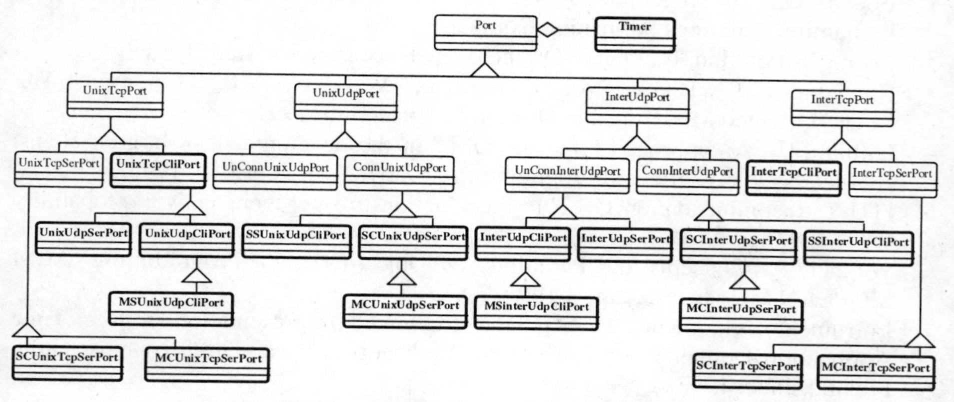

Fig. 13. A class library implementing port objects

7. Conclusions

In this paper an OO design method called ODM is presented. ODM is not a new OO radical method but it can be easily integrated with the most OO analysis methods. A new approach and some new quality concepts disregarded or hazy in other OO design methods are introduced by ODM. An ODM important characteristic is that before starting detailed design of a subsystem object, its quality profile has to be available as it is not possible to engineer quality before its specification. To define a quality profile ODM relies on ISO standard 9126 or on some other quality methods [11]. ODM introduces also a reference structure through which a systematic transition from specification to design is possible. This aspect is important as it fills, or reduces, the gap between analysis and design especially when dealing with either concurrent or distributed systems. Other ODM characteristics contributing to a better software quality are related to the support for many software process models, the management of system complexity, the improvement of system portability and maintainability.

ODM has been used for the OO design of the UPT service in the context of the IN and many positive responses have been gained. Future work is related to make some more experimentations of ODM for the design of software for real domains as well as to design and build a prototype of a C.A.S.E. tool for automating OSM and ODM activities.

Acknowledgements

This work has been accomplished thanks to the work of Eleonora Marchegiani for the specification and design of the UPT service. Special thanks go to Giuliano Paris for the documentation related to the UPT service and to Daniela Costanzi for her precious help.

References

1. Booch, G., Object Oriented Analysis and Design with Applications. The Benjamin/Cummings Publishing Company, 1994.
2. Coad, P., Yourdon, E., Object-Oriented Analysis. Prentice-Hall, 1991.
3. Rumbaugh, J., Blaha, M., Premerlani, W., Eddy, F., Lorensen, W., Object-Oriented Modeling and Design. Prentice-Hall, 1991.
4. Gilliam, C., An approach for using OMT in the development of large systems. Journal of Object-Oriented Programming, Vol. 7, N. 9, February, 1995.
5. ITU-T Recommendation Q.1211: Introduction to intelligent network capability set 1. 1993.
6. Wegner, P., Concepts and Paradigms of Object-Oriented Programming. ACM SIGPLAN Messenger, August 1990, 7-87.
7. Lofrumento, G., Fazio, V., An Object-Oriented Representation of Real-Time Application Domains. 5th Euromicro Workshop on Real-Time Systems, Oulu, Finland, June, 1993.
8. CCITT Recommendation Z.100: Specification and Description Language. Blue Book, Vol. X.1 - X.5, 1988.
9. Monarchi, D., E., Puhr, G., I., A Research Typology for Object-Oriented Analysis an Design. Communications of the ACM. Vol. 35, n. 9, September 1992.
10. Robinson, P., J., Hierarchical Object-Oriented Design. Prentice-Hall, 1992.
11. Deutsch, M., S., Willis, R., R., Software Quality Engineering - A Total Technical and Management Approach. Prentice-Hall, 1988.
12. Lofrumento, G., Objects Specification Model. SSGRR internal technical report n. 2961, 1994.
13. ISO/IEC International Standard 9126: Information technology - Software product evaluation - Quality characteristics and guidelines for their use. 1991.
14. Embley, D., W., Kurtz, B. D., Woodfield, S., N., Object-Oriented Analysis: A Model Driven Approach. Yourdon Press, 1992.
15. ETSI, UPT phase 1: Universal Personal Telecommunications (UPT) Requirements on Information Flows and Protocols. TCTR NA-71301 version 0.2.0, 1993.

Development of a Model for Reusability Assessment

Tor Stålhane, Ph.D.,

SINTEF - DELAB, Trondheim, Norway

Abstract. This paper shows how the REBOOT project constructed a reusability assessment model, based on a survey of what software engineers consider important for reuse. The main result of this presentation is that software engineers agree on a set of characteristics for a reusable software component. These characteristics can be combined into a factor-criteria-metrics assessment model. Since the assessment model represents the views of the software engineers it has a high probability of being accepted in the industry.

1 Introduction

The work reported here was done at an early stage of the ESPRIT project REBOOT (REuse By Object-Oriented Techniques). The focus of the REBOOT project is planned reuse, i.e. software reuse needs a database containing a library of high quality, highly reusable components, developed for reuse.

In our opinion, one of the main reasons why many software reuse projects have failed in the past, was that they only collected a large amount of components, made them available through a catalog or a database, and then said to the developers: "See if you can find something useful".

The REBOOT starting point is that reuse must start with a catalogue of components that are assessed to be reusable and have high quality. The first problem that confronted us was thus how to assess the reusability of a component.

A natural starting point when looking for ways to assess reusability is to search in the available literature. When we did this, however, we found that their advises were both imprecise and contradictory. Our literature survey, which covered 23 papers on software reuse, is reported in [3].

2 The Method

Instead of using the results from the surveyed literature, we decided that it would be more profitable to survey the opinions of software engineers. We reasoned as follows:

1. The decision whether to reuse or write from scratch is taken by the individual programmer, software engineer or project manager, based on his confidence in the component(s) offered for reuse
2. This decision is based on the information available at the time of decision
3. It follows from points 1) and 2) is that a person can be asked how a certain characteristic influence his decision on whether to reuse the component or not.

As result of this, we designed a questionnaire. The questions in the questionnaire were based partly on the metrics suggested in the surveyed literature and partly on the experience of the software engineers at SINTEF - DELAB and Sema SAE, which were the two REBOOT partners responsible for the reusability assessment. The questionnaire was distributed to several software engineers for assessment. In particular we wanted answers to the following questions:

- Are the questions in the questionnaire meaningful?
- Are there software characteristics that influence your reusability assessment which are not included?
- Is it easy to complete the questionnaire?

The input we got from this trial run resulted in a new version of the questionnaire. This version was distributed to software development companies in Norway, Sweden, Germany, France and Spain.

In order to simplify the assessment process, we wanted the software engineers to focus on "As is" reusability. Thus, all reusability assessment in this paper should be qualified with the condition "Given that the component performs the needed functions".

All the companies completed at least one questionnaire, some of them distributed the questionnaire to several software engineers in order to get a larger response set. All, in all, we received 29 filled-in questionnaires which were later used to obtain the reusability assessment model.

3 The Questionnaire

We had decided to use a qualitative method based on categories. The following schema was selected for assigning scores

++ Strong positive influence on reusability
+ Some positive influence on reusability
0 No influence on reusability
- Some negative influence on reusability
-- Strong negative influence on reusability

In addition to questions concerning reusability, we also collected data on each respondent and what kind(s) or reuse that where important in his daily work. For these two chapters of the questionnaire, see [4]. The rest of the questionnaire is structured as shown in figure 1.

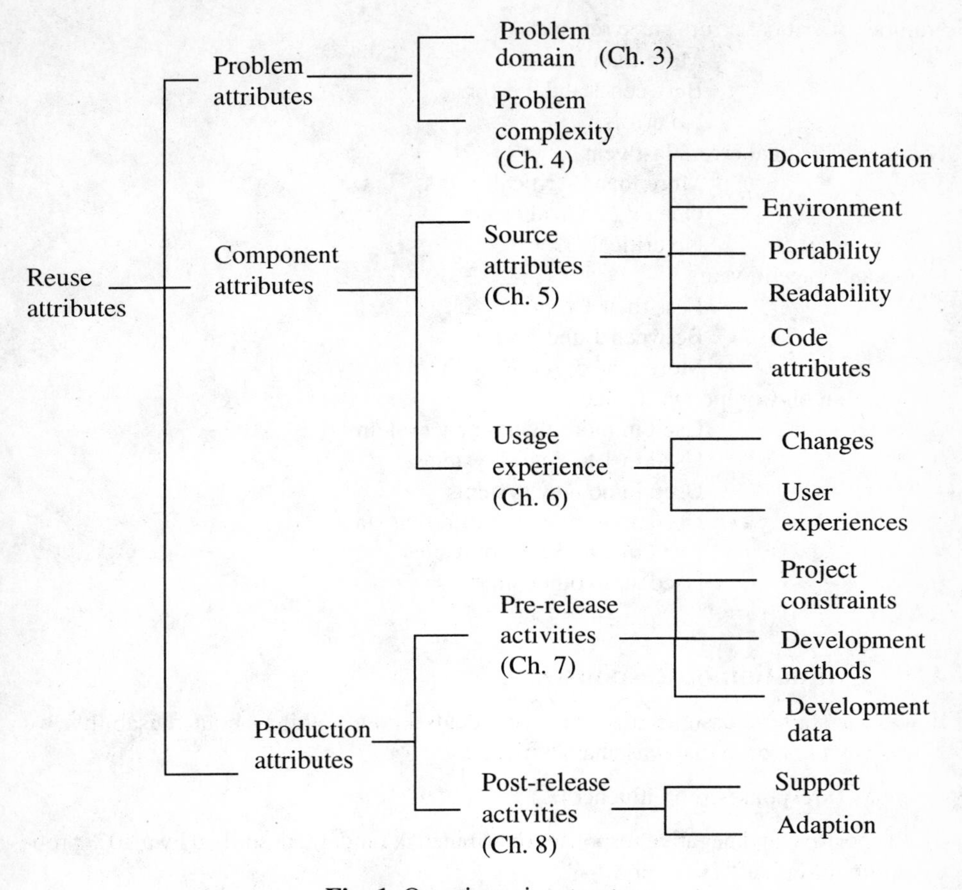

Fig. 1. Questionnaire structure

Each chapter in the questionnaire was split up into two or more subchapters as shown above. E.g. "Usage Experience" - chapter 6 - is split up into "Changes", chapter 6.1 and "User experience", chapter 6.2.

In addition, each chapter had a subchapter called "Other factors you consider important". This part of the questionnaire was, however, almost never filled in by the respondents.

Each subchapter had a set of topics which again were split up into a set of related characteristics. As an example we have included the split-up of "User experience", chapter 6.2 in the questionnaire.

Information on component usage

Mainly used in production environments
Mainly used in space or military systems
Mainly used in research and development
Mainly used in universities

Number of critical errors since release

More than 3 errors

Between 1 and 3 errors

No errors

Number of critical errors last year

More than 2 critical errors

One or 2 critical errors

No critical errors

Component age in years

Less than 1 year

Between 1 and 5 year

More than 5 years

Information on Component Usage

Used in more than 5 new systems

Used in 1 to 4 new systems

Used in no new systems

Used at more than 5 different sites

Used in 1 to 5 different sites

Used at no other sites

4 Evaluation of Responses

If we, for starters, assume that the respondents know nothing about reusability, we would get a response that was characterized by

- many 0-responses (no influence)
- the positive and negative responses distributed at random, described by a 50% probability of either "+/++" or "-/--".

We will only consider a characteristic as relevant for reusability assessment if we can reject the Ho-hypothesis stated as follows

Ho: There is a 50% probability of giving a positive (or negative) response.

- Let the total number of respondents be N
- Let the number of respondents who rank a particular characteristic as irrelevant (score 0) be No.

We reject Ho at the α-level if

$$\sum_{i=o}^{n+N_o} \binom{N}{i} \frac{1}{2^N} < \alpha \qquad \text{where} \qquad n = \min(N+, N-)$$

In this paper, we have decided to use $\alpha = 0.05$. This gives us the following significant reuse characteristic:

$$\sum_{i=o}^{N_o+n} \binom{N}{i} \frac{1}{2^N} < 0.05 \qquad \text{where} \qquad N = 23 \Rightarrow N_0 + n < 7$$

Thus, we will accept that the score is significant if we have N+ > 16 (positive influence) or N- > 16 (negative influence). Otherwise, we will have no significant influence.

By using this evaluation method and assigning the majority score (++, +, - or --) to each characteristic, we identified 31 significant characteristics. These characteristics are shown in full in appendix A.

When looking at the scores we discovered right away that

- there were no differences between the scores from different countries, even as far apart geographically as Norway and Spain
- some of the respondents had an academic background, while others had been educated in technical colleges. This difference did not show through.

Thus, it seems that we have really captured the characteristics that are important for software engineers for reusability assessment.

Some of the scores contain strong opinions (++/--) while others are somewhat weak (+/-). When we later refer to the questions by their numbers, we will enclose the weak response in parenthesis, e.g. "(5)".

5 A Reusability Assessment Model

The results from the questionnaire evaluation can be used to build a reusability assessment model. In order to simplify the process, we made the following decisions:

1. We would use the Factor-Criteria-Metrics model, as defined by [2]
2. Since the questions in the questionnaire mostly are concerned with low level, software-near characteristics, we would build our model bottom up.
3. In order to be able to compare our results to others, we would try to use as much as possible of the standard criteria and factor definitions. We decided to use the terms defined in the IEEE Glossary, [1]

The 31 questions that were found to be important for reuse can be grouped as follows:

- Easy to understand the component, both externally and internally. This includes 15 questions, which is circa 48% of all significant characteristics.
- Confidence in the component. This includes both reliability, robustness and that the component - if it fails - should not have adverse effects on the rest of the system. This includes four questions, which is circa 13% of all significant characteristics.
- Ease of porting the component to another hardware and/or software environment. This includes two questions, which is only 6% of the significant characteristics.

- Component history information, which includes development and user experience. This includes ten questions, which is circa 33% of all significant characteristics. We will later include this in the confidence-related information.

Thus, our results can be summed up as follows: "A reusable component must be easy to understand, we must have confidence in it and it must be easy to port it to another environment". In our opinion, the confidence part is too often left out of considerations.

6 Mapping Characteristics onto the FCM Model

We can take each of the groups of characteristics, namely Understandability, Confidence and Portability, and relate all the characteristics in each group to the Factor-Criteria-Metrics - FCM - model. Each quaesitum / characteristic has as irs answer one or more metrics values.

Understandability: The respondents separates the information of the component into three criteria:

- Documentation and interface information. This is information that can be found without having to look at the code. It is a black box view of the component.
- How easy is it to read the code?
- How complex is the code - or maybe more appropriate - how complex is the algorithm that we have used to solve the problem at hand?

Confidence: The respondents use four criteria to describe how they build up confidence in the component, namely

- Robustness.
- Maturity.
- Development process history.
- Use experience.

Portability: The respondents considered the following criteria to be important

- Machine (hw) independence.
- Compiler and system independence - later called implementation dependency.

It was decided right from the start that we wanted the FCM model to be as small as possible. In order to reduce the number of metrics, we decided to ignore all weak characteristics - characteristics that have receive a + or - score - except if this lead to the removal of one or more criteria. If we combine all the criteria above, based on this decision, we can build the FCM model shown in fig. 2 below.

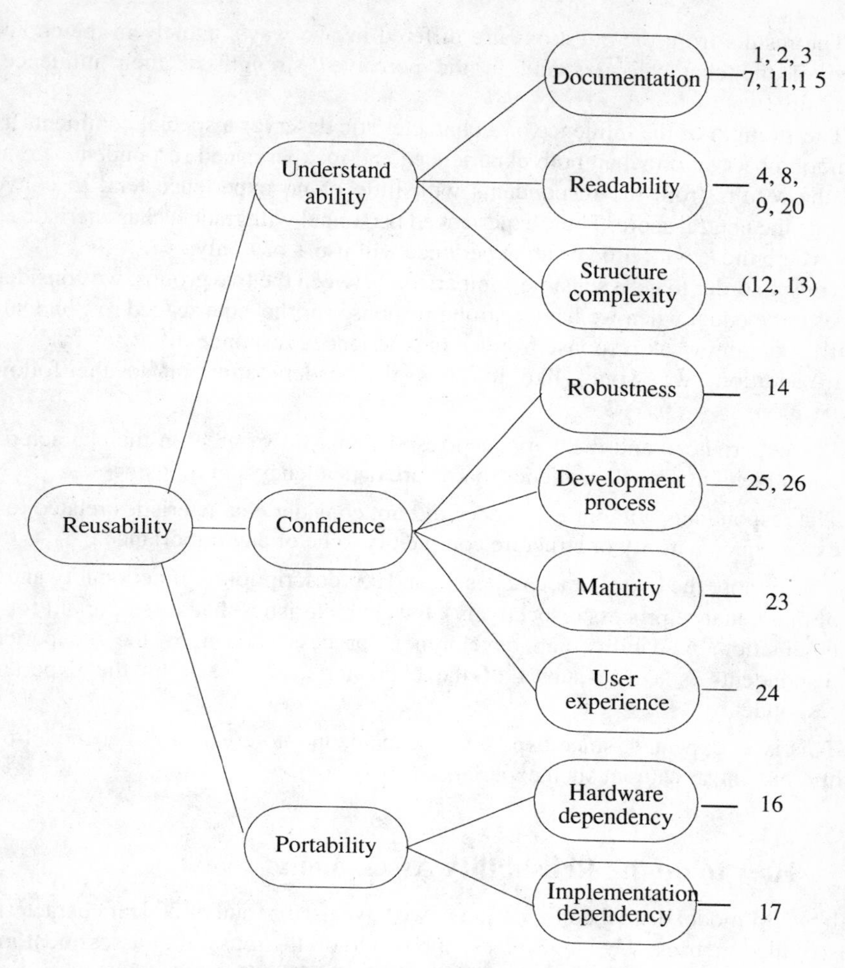

Fig. 2. The REBOOT model for reusability assessment

The metrics are indicated by their related questions numbers only. For a complete explanation, please see the characteristics listed in appendix A.

7 Responses from Personnel with no Experience

In addition to 23 responses from personnel that had been involved in one or more projects where reuse was applied, we got responses from six persons who had no reuse experience. We decided to analyze these responses separately to see if there were any differences between the two groups.

The results from the questionnaire differed in two ways, namely in the choice of some of the characteristics and in the perceived strength of their influence on reusability.

The strength of the influence of a characteristic deserves a special comment. It is a common observation when both experienced and inexperienced respondents are used, that the results from the respondents with little or no experience tend to converge towards the neutral score. Thus, experienced personnel will grade a characteristic as ++ or -- where those with little or no experience will use + or - only.

Thus, in order to get a sensible comparison between the two groups, we consider the scores to be equal when we have a strong response for the experienced respondent and a corresponding weak response from an inexperienced respondent.

In addition, we would like to draw the readers attention to the following observations

- The experienced and inexperienced respondents fully agree on the characteristics for implementation dependency, hardware dependency and robustness.
- The respondents without experience did not consider characteristics related to user experience, maturity or structure complexity to be of any importance.
- If we ignore the two characteristics 32 and 33 (description of functionality and type of maintenance process respectively), the characteristics that are important for documentation, readability and development process chosen by the inexperienced respondents is a true subset of the characteristics chosen by the experienced respondents.

This is as expected, since experience leads people to create a richer model of the reality than those without such experienced.

8 How to do the Reusability Assessment

In the FCM model described in section 6, we have used a total of 20 leaf characteristics - also called metrics. We can use these metrics to do the reusability assessment in two ways, namely qualitatively and quantitatively. In REBOOT we decide to use a quantitative assessment model only.

8.1 Assessment Requirements

First, we need a set of requirements for an assessment method. In order to assess the reusability of a software component for the REBOOT database we need at least to be able to rank a component on a reusability scale. This ranking must be done in such a way that it meets the following minimum requirements:

- The ranking must be "public". This implies that it must be clear to the reader how an assessor has arrived at his conclusion. "Public" does not necessarily mean "objective"; the important thing is that it must be possible to explain why the component has gotten its present ranking.

- It must be possible to use the ranking rules as guidelines during design and coding of a component. Thus, an example of a good rule is "No component shall have more than 100 lines of code", while a corresponding bad rule is "All components shall be small".
- If a component is rejected, it must be possible to deduce a set of improvement steps from the assessment. By following these steps we are guarantied to improve the ranking of the component as long as the same ranking method is used.
- The ranking must be considered fair. This implies that any competent software engineer must be able to recognize that the reasons for ranking a component as for instance highly reusable, really are relevant to reusability as he understands it.

Thus, we needed a set of software metrics, where each metric should measure one of the characteristics that influences a software component's reusability. The component's reusability can then be estimated as a weighted sum of these measures.

8.2 Quantitative Reusability Assessment

We decided that all metrics, criteria and factors should be computed in such a way that they always yield a value between 0 and 1 - the limits included. Furthermore, we decided that the factors should be weighed sums of the criteria and the criteria should be weighted sums of the measurements - metrics.

The results of the measurements should be understood within the framework of success probabilities. E.g. If a measure yields the score of 0.3 it does not necessarily imply that the component is bad; it does, however, indicate that reuse activities depending on this characteristic may run into problems - i.e. that the success probability is low.

When we implemented the reusability assessment model as part of the REBOOT qualification tool, we split all metrics into two groups - those that can be extracted from the software and those that had to be extracted via checklists. We will just show one example of each. A complete set of definitions can be found in [5].

- Robustness was decided based on two checklist questions, here denoted as Q1 and Q2. If the component contains several procedures, the component's robustness is computed as the straight average of the procedure robustness. The questions used have to be answered with a Yes - score 1 - or No - score 0. The robustness is then estimated as:

$$\text{Robustness} = \left(\sum \frac{Q1_i + Q2_i}{2}\right) / \text{Number of procedures}$$

- Implementation dependency is computed from information taken form the parser. The parser was augmented with a list of system dependent constructions and the code lines that contained such constructions were flagged. The implementation dependency is then estimated as:

$$\text{Implementation dependency} = 1 - \frac{\text{Number of system-dependent constructions}}{\text{lines of code}}$$

In the same way, all the metrics needed where computed. The reusability score is defined as

$$\text{Reusability} = \left(\sum (w_i \times \text{Metric}_i)\right) / \left(\sum w_j\right)$$

The weight w_i describes the importance of each metric. Since the importance of a metric can change from site to site, the REBOOT tool allows the person responsible for the reuse library to change these weights if needed. An example of such a need is if we have decided not to port the components to another platform. In this case, it is reasonable to set the weights for hardware and implementation dependencies to zero.

9 Conclusions

Metrics models have often failed due to lack of credibility. Our assessment model represents the view of experienced software engineers. This gives the model high credibility and will increase the model's probability of being accepted in an industrial environment.

Our main conclusions are as follows:

- Software engineers spread across Europe agree on what characterizes a reusable software component.
- Reusability is a complex relationship and can not be assessed by a few, simple metrics like v(G) and lines of code.
- The metrics that are important for software reusability are not language dependent but contains a set of general characteristics usually collected under the headings "Good programming style" and "Confidence building".

Reference

1. IEEE Standard Glossary of Software Engineering Terminology ANSI /I EEE Std 729-1983

2. IEEE Standard for a Software Quality Metrics Methodology.Draft standard P-1061 D21, 1990-04-01

3. Tor Stålhane and Eldfrid Ø. Øvstedal, Literature Study of metrics for Reusability REBOOT report 91-01-25

4. Tor Stålhane and Alfredo Coscolluela, Final Report on Metrics, REBOOT report 7090.1, 1992-02-12

5. Tor Stålhane and Alfredo Coscolluela, Functional Analysis for the REBOOT Metrics Tool, REBOOT report 7156.7, 1994-07-23

Appendix A Significant Characteristics

1. Documentation of the problem the component is supposed to solve

++	Formal description of components problem
+	Informal description of components problem
--	No description of component problem

2. Documentation of the problem the system is supposed to solve

++	Formal description of system's problem
+	Informal description of system's problem
--	No description of system's problem

3. Available component header information

++	Description of the problem solved
+	Description of the used algorithm
++	List of variables
+	Other components called
-	No header available

4. Documentation available as comments

+	Description of algorithm
+	Explanation of main variables
++	Description of each code block
+	Comments on each decision point
++	Comments on machine dependent code
+	Comments on bug fixes

5. Simple demonstration examples

+	Extensive set of demo programs available

6. External documentation available

+	Cross reference available
+	System's requirements specification
+	Component requirement specification

7. Interface description

++	Formal interface description
+	Informal interface description
--	No description available

8. Mnemonic constant names

++	Mnemonic names for all constants
+	Mnemonic names for key constants
--	No mnemonic constant names

9. Code layout

--	No formalized code layout rules
+	Code identification used for IF-THEN-ELSE, CASE alternatives etc.

10. Input and output section
    - ++ Input and output separate from code and from each others
    - - No separation
11. Structure diagrams
    - ++ Complete structure diagram available
    - + Simplified structure diagram
    - - No structure diagram
12. Number of functions provided
    - - More than five functions
13. Code complexity
    - - More than five nested levels
14. Code robustness
    - ++ All input parameters checked
    - - No input parameters checked
    - ++ All error exits documented
    - -- No error exits documented
    - ++ All errors controlled inside components
    - -- No errors controlled inside components
    - + Default values supplied for all output parameters
15. List of external dependencies
    - + List of all functions called
    - + Diagram showing the call hierarchy
    - -- No information available
16. Number of machine dependent LOCs
    - -- More than 50%
    - -- 25-50%
    - ++ No machine dependent code
17. Number of non-standard language LOCs used
    - -- More than 50%
    - -- 25-50%
    - ++ No non-standard language use
18. Standard test cases
    - + Large set of documented test cases
    - + Some documented test cases
19. Comments in the code
    - + Comments on each decision point
    - + Block of comments for each chunk of code
    - - No comments for each chunk of code

20. Mnemonic variable names
    - ++ Mnemonic names used for all variables
    - \+ Mnemonic names used for key variables
    - -- No mnemonic names used

21. Version and variant history
    - \+ Complete history supplied
    - \+ Information on major changes available

22. Size change since release due to corrections
    - \- More than 40% changed

23. Number of critical errors since release
    - -- More than three errors
    - ++ No errors

24. Information on component usage
    - ++ Used in more than five new systems
    - ++ Used at more than five different sites

25. Types of development constraints
    - ++ High portability requirements
    - \+ Requirements on testing strategy

26. Description of development process
    - ++ Strongly formalized development

27. Type of quality assurance used during development
    - \+ A documented QA standard

28. Development tools used
    - \+ Design tools
    - \+ System simulation tools
    - \+ CASE tools

29. Levels of component support
    - \+ Support available on call
    - \- No support available

30. Type of quality assurance used during maintenance
    - \+ A documented QA standard

31. Number of errors per change after release
    - \+ No errors found

The Impact of Reuse on Software Quality

Even-André Karlsson
Q-Labs, S-22370 Lund, Sweden
eak@q-labs.se

Jean-Marc Morel
Bull S.A., F-78340 Les Clayes-Sous-Bois, France
J.M.Morel@frcl.bull.fr

Abstract. This paper focuses on the impact of reuse on the various factors (functionality, reliability, usability, efficiency, maintainability, and portability) which make up the Quality of a software system. Then, knowing all the impacts reuse has on the Quality, we will present the approach we have developed in the REBOOT[1] project to help a software producing organisation adopt systematic reuse and so increase quality and productivity. But, the current acquisition procedures usually do not encourage reuse. The worst of it is that it often impedes reuse. Therefore, we propose guidelines to follow throughout the acquisition process to favour and organise reuse.

1 Introduction

Reuse is perceived as one of the major possibility to improve software productivity and quality in the future. Barry Boehm at the STARS conference in 1991 illustrated the relative importance of three sources of cost saving in software development (see Figure 1).

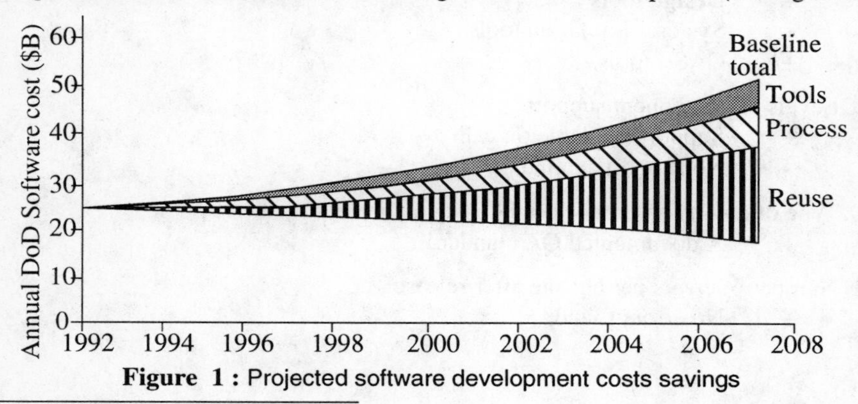

Figure 1 : Projected software development costs savings

1. REBOOT (REuse Based on Object Oriented Techniques) is an ESPRIT III project (EP7808) which started September 1990 and finished October 1995. The Partners were Bull S.A. (Coordinator - France), Cap Gemini Innovation (France), Sema-Group sae (Spain), Siemens AG (Germany), Q-Labs (Sweden), EP Frameworks (Sweden), SINTEF (Norway), and TXT Ingegneria Informatica SpA. (Italy).

Data supporting this prediction showing the trend in quality, cost and reuse from NSAS/ GSFC were presented by Frank McGarry at a seminar at the European Software Institute in October 1994 (Figure 2). Figure 2 shows that there is an improvement in quality (4

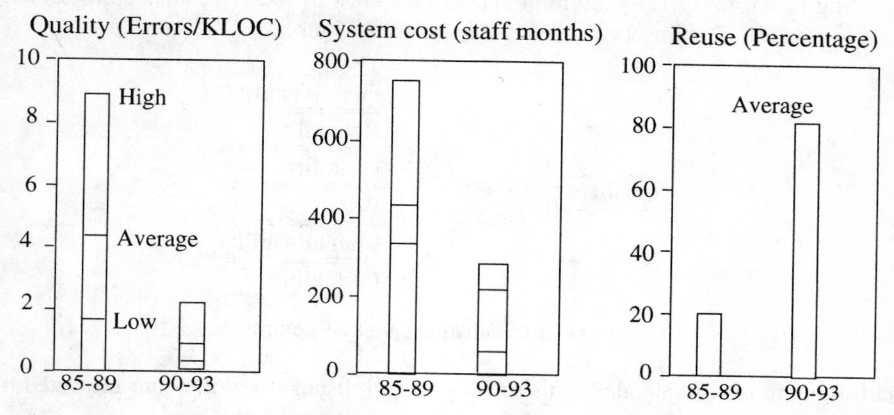

Figure 2 : NASA/GSFC quality, cost and reuse data

times) and productivity or cost (2 times) in connection with an increase in reuse. The data are collected from 7-8 similar systems in each period. This does not give any proof that reuse was the major contributor to these improvements, but other improvements, e.g. changes in tools (CASE) and process (Cleanroom), were introduced and measured in the same period, and neither gave more than limited improvements.

To understand how this effect of reuse on quality is achieved, and even more important how we can continue to enforce it we need a better understanding of how reuse affects quality. We have approached this by analysing the effect of reuse on the ISO 9126 definition of software quality. This is presented in Section 2.

Having understood how reuse can impact quality we need to analyse how we can encourage the production of more reusable components, as well as the reuse of these components. The analysis of what constitutes a reusable component and how to develop it is treated in Section 3.

Much software is developed under contract, and existing acquisition procedures do not encourage reuse, some even discourage it. In section 4 we discuss how reuse can be encouraged and incorporated into current acquisition procedures.

2 Reuse and quality

Quality is usually defined as the ability of software to meet its requirements. We distinguish here the different factors which are incorporated in the concept of quality according to the ISO 9126 definition [1], and for each of these we analyse the consequences of reuse. The observed factors are shown in Figure 3.

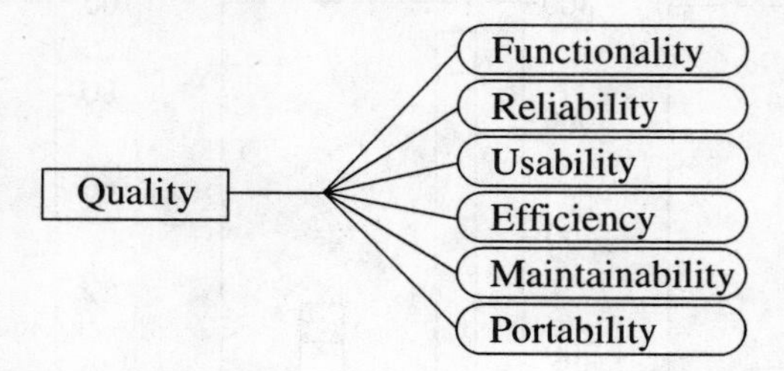

Figure 3 : Software Quality Factors

The following paragraphs define these factors, and discuss how they are impacted by reuse:

- **Functionality** is the existence of a set of functions and their specified properties. The functions are those that satisfy stated or implied needs. Functionality is relative to the customer requirements, but as most customers have unclear expectations in the initial phases of software development, reuse can affect these requirements in several beneficial ways:
  - Reuse provides an opportunity to reuse similar systems as prototypes to capture the functional requirements.
  - In development *with* reuse, we can uncover hidden requirements during the evaluation of potential reusable components. The evaluated components can also help the customer revise the requirements, a similar effect to that achieved with fast prototyping.
  - In reusing a component there is a risk that we sacrifice the "right" functionality in favour of reuse, but there may also be an opportunity to provide additional functionality which was not possible without reuse.
  - If we have an opportunity to reuse a system covering for example 95% of the customer requirements without modification, the additional 5% of the functionality is in most cases negotiable, provided the customer can share some of the benefits of reuse.
  - In developing a system *for* reuse there are usually several customers involved. They can influence each other and elucidate latent requirements which might not have emerged until later in a traditional development process, causing considerable problems.

- In development *for* reuse there is always a compromise between different customers for the component. Thus some conflicting functionality might have to be sacrificed, but additional functionalities provided for other customers may be gained.

The process of developing *for* and *with* reuse, and its consequences, are discussed in more detail in Section 3.

- **Reliability** is the software's ability to maintain its performance level under stated conditions for a given period of time. Reuse can influence reliability both in a positive and negative fashion:
  - The correct reuse of a well-tested component increases the reliability of the system (positive).
  - Several uses of the same component increase the confidence in the component (positive).
  - Incorrect reuse or reuse outside the intended scope of a component, however well-tested, constitutes a risk (negative).

  The importance of proper documentation and evaluation of components is further discussed in Section 3.

- **Usability** is the effort needed to use the system by the total set of users. Reuse also has impacts on this aspect of quality, both positive and negative:
  - A reused component usually has more effort put into its development than a one-off component (a positive effect on the usability of the component and the system).
  - Components with slightly different characteristics might be unacceptable from a usability point of view (negative). The evidence of this can be found in the market for administrative systems, where even with an abundance of cheap commercial standard packages, companies opt for custom development instead of changing their procedures to fit the available products.
- **Efficiency** is the relationship between the performance level of the software and the amount of resources used. This is also affected by reuse:
  - More effort can be justified to make a reusable component efficient than for a specific component which will only be used once.
  - Specific components can be tailored for specific needs, and thus be more efficient than a more general reusable component. The reuse of a component tailored for specific needs might downgrade the efficiency in other environments.
- **Maintainability** represents the effort needed to make specific modifications. For a reusable component, many of these modifications are already incorporated or planned in the extended requirements of development *for* reuse. Modifications outside the intended plan are more difficult, as the functionality of the component is more complex than a specialized component, but the cost of modifications of a reusable component can be divided between the various reusers.

- **Portability** is the ability to be transferred from one environment to another. This aspect of quality is directly correlated with the reusability of the component, and portability analysis is an important aspect of development *for* reuse. We can compromise the portability of the system by reusing components which are not portable.

We should always bear in mind that as for any quality improvement, the development of reusable components is not without cost. Perhaps the most important lesson to be learned in connection with reuse is that the development of reusable components should be treated as an investment, the profits from which are reaped by using them outside the product for which they were developed. In the next section we will investigate how we make components more reusable.

3 Development for reuse

What distinguishes a reusable component from a non-reusable one? A reusable component is developed in one context and is intended to be reused in other contexts. To be able to reuse a component, it must fulfil a reuser's need for a specific functionality. The clue is to foresee this need, and define a component so that as many reusers as possible can profit from using it, thus:

Reusability is useful generality

Even if non-functional criteria like reliability and maintainability are of major importance when it comes to reuse, it cannot be overlooked that the first and most important criteria for reuse is functionality. If a component does not solve our problem, we do not need it.

As different reusers usually have different requirements for the same component, it is not profitable to construct one "concrete" component for each reuser. It is better to construct general components that can be specialized, parameterized or configured by different reusers. This means that we need components that are as general as possible. We must, however, also consider the cost of understanding, adapting and integrating a general component, as well as any performance penalties. These factors indicate that we must carefully weigh the amount of generality we build into a component. The conflict is illustrated in the following simplified figure:

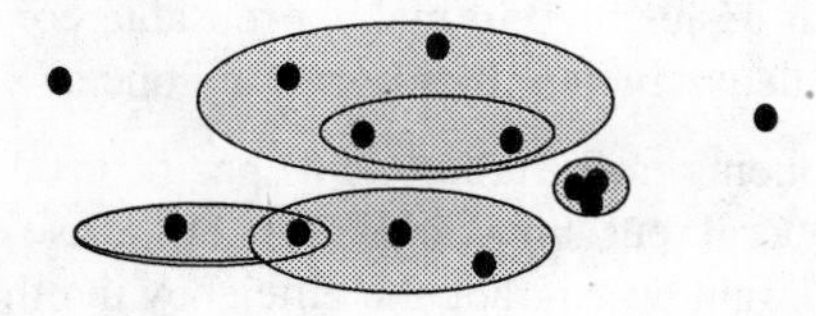

Figure 4 : General components and specific requirements

Here different reusers' specific requirements are illustrated with black points, the distance between the points represents the difference in requirements, and the reusable components are shaded ellipses, the larger ellipses representing more general components. The cost of evaluating, understanding and adapting a reusable component can be thought of as proportional to the cost of contracting one of the ellipses to the size of a specific point. We must therefore balance generality with the cost of adaptation. The only way to get this balance right is by making a thorough analysis of the current and potential requirements. A similar process is described as *instance-space analysis* in [2].

We defined development for reuse as the planned activity of constructing a component for reuse in contexts other than that for which it was initially intended. Development for reuse therefore extends and generalizes the requirements of the component for the current user, so that requirements for future reusers are captured, and the component can be reused.

Analysing requirements from potential customers leads to a better understanding of the original customer's requirements. This might well lead to early discovery of hidden or undiscovered requirements from the original customer, which would otherwise lead to expensive rework if discovered later in the development process.

3.1 How can you represent reusability?

When we decide which requirements it is economical to include in a reusable component, we must decide how best to represent the reusability or generality.

This is a design activity in which we must take into account what is technically a good solution, and whether potential reusers will be able to understand and adapt the component easily.

In analyzing several reusable components, we have found that it is possible to identify several general techniques which may be applied independently of component size and the development process step. We decided to focus on the following five techniques. Although we do not claim them to be complete or orthogonal, we have found them to be very useful in practice:

- *Widening*. This means identifying a set of requirements that are not contradictory, then making a general component that satisfies all of them. The advantage with this approach is that the component can be reused *as is*. The disadvantage can be the cost of the component, with respect to:
    - initial development, i.e. it is expensive to develop all the functionality at once, and the component may become unnecessarily complex,
    - being obliged to understand irrelevant functionalities when reusing the component,
    - resource consumption, e.g. space and efficiency.

- *Narrowing*. Here we identify functionality common to several customers which can be represented by an abstract component. We can choose to place an abstract component at any level in the requirements hierarchy. Consider the following abstraction hierarchy:

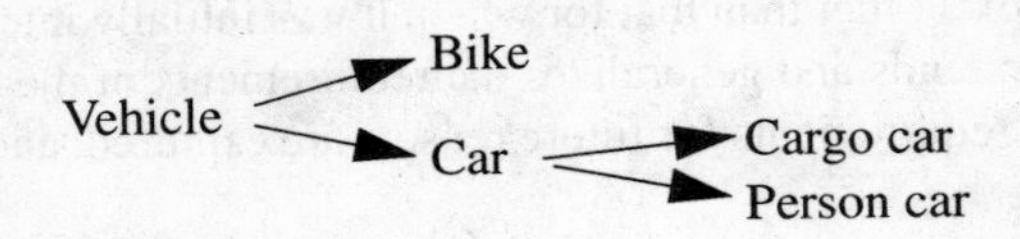

Figure 5 : An abstraction hierarchy

When we want to make a new vehicle, such as a monocycle or a convertible, we will specialize *Bike* or *Car* respectively, rather than implementing it from scratch. The principle of reuse by inheritance is well known within object-orientated development, where it is most commonly applied to classes. There is no reason why it should not be applicable to larger components such as modules, subsystems or entire systems.

- *Isolation*. Different requirements can be isolated to a small part of the system, and the rest of the system constructed relatively independently of whatever specialization is chosen. A special case of isolation is *parametrization*, where we recognize that some requirements can be expressed by parameters. Other examples are *indirection* and *abstract interfaces*, that is, using pointers to objects where we only specify the required interface, and leave the rest of the object unspecified. Layered architectures are also a special case, where we isolate variable parts in different layers. Isolation is a much-used technique in making components independent from the underlying system, for example the operating system, hardware or database server.

- *Configurability*. Configurability means that we make a set of smaller components, that can be configured or composed in different ways to satisfy different requirements. For the car we could imagine, for example, a modular middle part to the chassis to enable us to build different sizes of car (one or two middle modules respectively) and an engine with a variable number of cylinders. This is an even more viable approach for optional requirements.

  Configurability exists on a scale ranging from a strict framework, where we can only compose the system according to given rules (see [3] and [4]), to a general-purpose library that allows us to build systems from very general "bricks". This can be compared to the popular Lego toys, which provide everything from general-purpose kits to very specialized ones. Interesting work is going on to try and categorize object-oriented design patterns that can be reused [5]. These are the kinds of structures that can be used to achieve configurability.

- *Generators*. Different requirements can be satisfied by making a "new" application-domain-specific language with which one can describe an application. Executable code can then be automatically generated from the application description. This approach requires a mature knowledge of the application domain and a considerable investment.

Even with application generators we can use component-based reuse, because the programs for the application generator can be reused. By doing this we move the reuse problem one layer up.

These techniques are not orthogonal, as we can use more than one technique to represent the same generality. Which technique we use is a design decision. It is important, however, that designers are aware of these techniques, because they tend to appear in different guises, and a decision to use any of them will affect a component considerably.

For example, object-oriented framework technology is based mainly on a combination of the *configurability* and *narrowing* techniques, resulting in a set of abstract classes with predefined communication mechanisms.

3.2 How do you develop *for* reuse?

How should you develop reusable components? We have argued that development for reuse is independent of both component size and life cycle phase. We have also found that a component can have different degrees of generality. This means that we can try to identify the steps in a general development process for reuse independently of:

- where the component is in the system structure, for example an entire system or a single class
- what kind of development model we use to represent the requirements and solution

We list below the steps we have found necessary to follow in development for reuse. The resources we put into each of these steps depends on the size of the component:

1. Capture the initial requirements, collecting the set of requirements to make an initial solution.
2. Define an initial solution or identify previous solutions to the same set of requirements. It is best to keep the solution in a form that customers can understand and validate. In the case of re-engineering, the result of this step should be the identification of previous solutions. A cost estimation for developing the actual solution should be made or updated at this time.
3. Identify possible generalizations. This is the inventive step in which we try to see the generality in our requirements and our solution, based on both the requirements and the initial solution. At this step the component's reuse potential should be discussed with the product management and the plan for following steps should be refined.
4. Identify potential reusers and collect their requirements. Potential reusers are important to ensure that the generality we include in our component is really justified.
5. Estimate the cost and benefit of added functionality. For each added requirement we must estimate:
   - the benefit for reusers who will use it,
   - the extra cost (e.g. time to understand, remove) for reusers who do not need it.

   Each potential reuser should also estimate the effort needed to develop the functionality he needs from scratch, and the probability that he will reuse the component.

6. Analyse the added requirements with respect to invariants and variation. This analysis usually impacts the initial solution, and we should therefore be prepared to modify it if necessary.
7. Propose a generalized solution with specializations and cost estimates. Here we reconcile the different requirements into one solution by applying techniques to represent generality from the appropriate models.
8. Present the solution to reusers and reuse experts for validation and approval. Each reuser must ensure that the solution covers his requirements. He should also refine the estimate of how much it will cost him to reuse the component. The reuse expert's role is to ensure that the solution is satisfactory, and to check that the proposed specializations are appropriate.

   Based on this input, we can decide if we will develop the component for reuse. All the estimates made during the analysis of requirements and solutions should be saved to enable comparison with actual data. Note that this step is taken before we start the bulk of the development effort.
9. Develop and document the solution. This is the final step where we implement the component or system. If we are developing a high-level component we may repeat this process for reusable subcomponents. We discuss the documentation needed separately in Section 3.3.

These steps are generic and can be applied at any stage in the development process and system structure. They do not necessarily form a strict sequence: we may iterate over steps 4 to 8 several times.

We frequently observe that this more elaborate analysis process can be profitable even if we choose not to develop the component for reuse. For example the search for a more general solution and the study of other potential reusers can uncover hidden requirements from the original customer, lead to a more complete and adequate solution, and avoid costly changes later in the development or maintenance phase when problems arise due to missing functionality.

3.3 Documenting a reusable component

Documentation is the glue between the producer and the consumer of the reusable component. It is an important means for the producer to communicate the purpose of the component to the consumer. Documentation must be tailored to the process of development with reuse.

In system development we can distinguish between different forms of documentation:

- Engineering documentation (such as project plans, test plans, and so on). This is documentation produced during all development stages from analysis to test. This documentation is usually not part of the product documentation.
- Product documentation. This is documentation accompanying the product that contains information for those who will use the product after its release, such as users, system administrators, sales staff, and installers.

- Maintenance documentation which contains information (such as requirements and design specifications) for those who will maintain or evolve the product.

To enable reuse, some of the maintenance documentation needs to be converted to product documentation. Engineering documentation should also be made accessible so that later projects can learn from the experiences of those who built the product. When we introduce reuse the distinction between these three forms of documentation decreases.

From an analysis of the different steps of development with reuse process, we can understand the need for specific reuse documentation better:

1. For *searching*. The components in the reuse repository must be organized so that a set of candidate components can be easily retrieved. This can be anything from an advanced faceted based classification to data sheets in a binder.
2. For *evaluation*. When a set of candidate components is retrieved we need to evaluate them to select one that we will try to reuse. For this purpose we need evaluation information. This should consist of one–two pages that describe the component and give an overview of its functionality.
3. For *investigation*. When we select a component we intend to reuse, we need to understand how to use it. This is like a user reference manual for the component, which can helpfully incorporate examples of use.
4. For *adaptation*. Many components developed for reuse are intended to be adapted before they are reused. The different approaches for representing this adaptable generality were described in Section 3.1, and should be reflected in the documentation:
   - *Widening* does not constitute any problems in documentation.
   - *Narrowing* relies on object-oriented concepts like inheritance and dynamic binding. We need to document the functionality of the abstract component, and then describe how it is supposed to be specialized.
   - *Isolation* requires a combination of *as is* documentation and specific documentation for what needs to be provided.
   - *Configurability* requires that we document the key mechanisms used to configure the components, that is, how the components are intended to be connected and communicate with each other, and which components can be connected together.

   Documentation for planned adaptations is the most important for reuse at this stage. A component can also be modified more radically than the planned adaptations, if they prove insufficient for a reuser's needs. In such a case, an entire set of maintenance documentation is required to support this activity.
5. For *integration*. Here we need to incorporate the documentation of the reused component into our existing system. The reused component is going to be part of a system, and needs to be documented as such. When we develop the component for reuse by adaptation there are two alternatives for the product documentation:

- Prepare the documentation for incorporation in the system documentation. This means that the total life cycle documentation from analysis to test should be prepared for this adaptation, so that the component can be incorporated into the new application in the same way as a component developed from scratch. This is the traditional way to document maintenance and enhancement of existing systems.
- Prepare the documentation so that only the adaptations are documented. This means that we can reuse the general documentation *as is*, and add our adaptations as separate documentation.

The strategies to use will be determined by the type and scope of the reusable component. A component to be reused once with minor adaptations in a similar application will be reused more readily if the documentation of the adaptations is incorporated. Documenting just the adaptations is more suitable for a generally reusable component. The special object-oriented mechanisms of frameworks and inheritance fall naturally into the latter case.

It is also important to include any relevant negative information with the component, i.e. what lies outside the intended scope of the component. This information is acquired from the steps in point 5 and point 8 of the development for reuse process described in Section 3.2.

3.4 How do you test a reusable component?

Testing a reusable component has to be performed both by the producer and reuser of the component. The producer must check that the component is as adaptable as intended, and test the correctness of the component's general functionality. The reuser has to test the adapted and integrated component. The effort needed to reuse the component will be reduced if good test documentation is provided with the component.

For reuse *as is* the testing provided by the producer can be quite comprehensive. The component should be provided with a large set of test cases. Guidelines for integration tests should also be provided. The reuser should assure himself that this set adequately covers his use of the component, and optionally make and test additional cases.

For reuse with adaptations the problem of providing test support for the reuser is more difficult. Ideally we should prepare tests for the adaptations as we do for the component itself, as described in Section 3.1. Some ideas on how to prepare tests for different techniques are:

- *Widening*. Tests can be prepared as for a component reused *as is*.
- *Narrowing*. Tests can be prepared by allocating the general test cases to general components. How to provide abstract test cases which can be specialized together with the component should also be investigated.
- *Isolation* is similar to narrowing, where just the test cases for the adaptations are isolated.

- *Configurability* represents a challenge when it comes to testing, because it is difficult to deduce suitable system tests from test cases for individual components. The best solution is to provide guidelines and standards for testing systems that contain the component. These guidelines should include testing of any communication mechanisms the component provides.

3.5 Measuring faults in reused components

Faults are not only failures during execution, but all kind of corrections required to finished products to make them conform to customer specifications. These might result from errors in requirements, design, or code, as well as documentation. Faults can be discovered during any kind of quality assurance or development activity. Faults can be traced back to errors in the code. Three types of errors in the reused code can be identified:

1. Errors in the original component.
2. Errors in the code written to modify the component.
3. Errors in the code using the component, caused by the use of the reusable component.

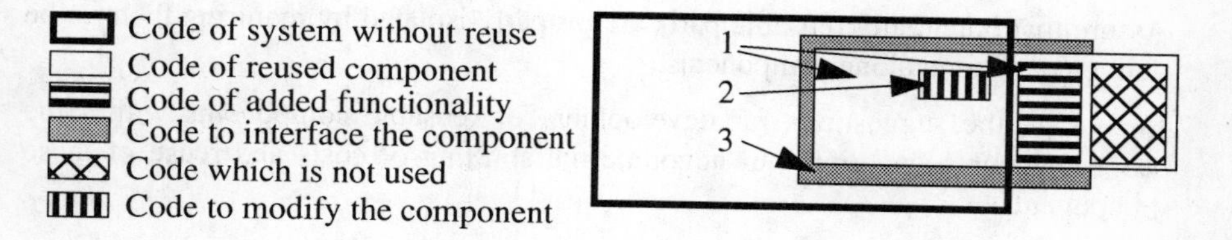

Figure 6 : Faults in code involved in development with reuse

These three types of code where faults can occur are shown in figure 6—we do not distinguish whether the reused code was needed, included or not used.

Faults in the code written to modify the component and faults in the code using the component are used to evaluate the reuse quality of the component. Based on these faults it should be easy to suggest improvements to the component or its documentation, and thus to the process producing or reusing such components, so that they can be avoided in future.

Since we use an estimate of the equivalent number of lines of code, measured quality will be relative to the functionality that we reuse. This gives a good picture of the total quality of the component if we assume that faults are evenly distributed. If some reusers experience a different quality than others, it may be that they have estimated the amount of functionality that they reuse in a wrong way, or that they are reusing the component in a different way to other reusers.

4 Software acquisition and reuse

Customized software is very often developed by software houses based on an acquisition process. To improve reuse in these situations we will look at simple guidelines which the acquiring party can apply to improve both development *for* and *with* reuse.

Preparing the Acquisition Plan

- Investigate existing reuse libraries that the contractor can use, their contents, their cost, and their availability.
- Analyse risks: What companies have experience in the domain and are able to implement reuse? Are there enough reusable components available to require reuse?
- Contract: Is there a need for a contractual incentive award to encourage reuse? How much reuse can be expected? How many new reusable components should be developed? Which available components need to be made reusable?
- Acceptance: Will reuse modify the acceptance procedure and the evaluation criteria? Will reuse reduce the evaluation costs?

Encouraging development *for* reuse

- Require the provider to describe how the system can be extended with possible future requirements.
- Assure that potentially reusable parts are properly isolated by requiring them to be delivered as stand alone components.
- Encourage the suggestions for development of reusable components within the scope of your acquisition, but negotiate the splitting of costs and reuse of these components.
- Check documentation of system parts with respect to reusability.

Encouraging development *with* reuse

- Ensure that reuse activities (searching, evaluating reusable components) are explicit in the development process and well scheduled.
- Provide your set of reusable components as part of the specification, allowing the reusers to make the fullest use of these in their offer.
- Ensure that adequate resources are available to execute the reuse effort (e.g. training, skills) and that responsibilities (roles) are adequately assigned.
- Support the reuser in understanding the reusable components if possible.
- Measure costs (searching, understanding, integrating) and savings (compare to development from scratch). This should take into account all steps from requirements analysis to maintenance. This will help in the writing other contracts and in convincing other people to adopt reuse.
- Ask people to fill reuse reports in order to help future reusers make decision, evaluate the cost of reuse, and adapt to their needs.

5 Conclusion

It is clear that reuse has an important impact on the quality of software products. We have seen that most of the quality aspects (functionality, reliability, usability, efficiency, maintainability and portability) are improved when reuse is implemented. This is no surprise but most software producing organisations still have ad-hoc reuse practice. Reuse must be organized and integrated with software engineering, that is reuse must evolve from being opportunistic to being systematic. We have given some hints and guidelines to help go in this way. A full treatment of the REBOOT approach to organized reuse is given in [6]. The main conclusion is that reuse must be taken into account in all aspects (acquisition, organisation, management, engineering, and technological) in order to be successful.

6 References

1. *Information technology - Software product evaluation - Quality characteristics and guidelines for their use*, Draft International Standard ISO/IEC DIS 9126, UDC 681.3.06.006.83.
2. Bruce H. Barnes and Terry B. Bollinger, *Making Reuse Cost-Effective*, IEEE Software, January 1991, pp. 13-24.
3. MacApp: The Expandable Macintosh Application version 2.0B9, Apple Computers 1989, Cupertino, CA, USA
4. Roy H. Campell, Nayeem Islam, Peter Madany, *Choices, Frameworks and Refinement*, Computing Systems, Vol. 5, No. 3, Summer 1992, pp. 217-257
5. Erich Gamma, Richard Helm, Ralph Johnson, John Vlissides, Design Patterns: *Abstraction and Reuse of Object-Oriented Design*, ECOOP 93, pp. 406-431
6. Even-André Karlsson (Ed.), *Software Reuse - A Holistic Appraoch*, John Wiley and Sons, forthcomming May 1995.

Practical Guidelines for Ada Reuse in an Industrial Environment

I. Sommerville[1], L. Masera[2], C. Demaria[3]

[1]Computing Dept., Lancaster University, LANCASTER LA1 4YR, UK.
E-mail: is@comp.lancs.ac.uk
[2]SIA, Via Servais, 25, Torino, 10146, Italy.
E-mail: masera@sia-av.it
[3]ALENIA, Corso Marche, 41, Torino, 10146, Italy.
E-mail: APPRAISAL@test.alenia.polito.it

Abstract. An essential pre-requisite for widespread software reuse is a significant base of reusable components at different levels of abstraction. The construction of such a component base is simplified if components are developed, in the first place, to be as reusable as possible. Although there have been a number of studies of guidelines for producing reusable Ada components, we have found that these are too vague and general for practical application in an industrial setting. This paper will discuss part of the work going on in a technology transfer (ESSI) project called APPRAISAL. We are addressing the need for simple, practical reuse guidelines which can be applied without significantly increasing development costs. We describe some of these guidelines and the rationale for their development. We discuss requirements for reuse information dissemination and explain how World-Wide Web (WWW) browsers satisfy these requirements.

1 Introduction

The subject of software reuse is one which has been of interest to the software engineering community since the term 'software engineering' was coined in the late 1960s. In an early software engineering conference, McIlroy [1] proposed his vision of system construction from reusable components. Throughout the 1980s, various research projects addressed reuse issues. Amongst larger companies, at least, there is now an awareness that reuse has the potential to improve software quality and reduce software costs. In the US, a major reuse initiative is underway with over $150 million made available to selected companies to develop generic software and to tackle some of the complex intellectual property issues of software reuse [2].

Although awareness of reuse has increased significantly over the past few years, there still remain practical problems in introducing reuse into software development processes. We cannot discuss all of these problems here but the problems which are most relevant to the subject of this paper are:

1. How can an organisation acquire a base of reusable components at a reasonable cost?

2. How can engineers involved in development be encouraged to develop reusable components?

3. What are the trade-offs to be made between reuse, efficiency, reliability and other non-functional concerns?

There are various approaches which can be adopted to address these problems. The approach discussed here is based on the production of a set of guidelines for Ada programmers. These provide advice on how to improve component reusability. As we discuss later, the key characteristic of these guidelines is that they should be simple to apply and understand and as unambiguous as possible.

The work described here is being carried out in the context of the APPRAISAL project which is a technology transfer project, partially supported by the European Commission. The project is based around an application experiment in the domain of real-time, embedded aerospace systems. The objectives of the experiment are to combine reuse with safety and efficiency concerns and to select appropriate reuse methods according to their maturity and development process. Specific objectives include the development of a way of identifying and re-engineering existing software components (the principal topic of this paper), and the investigation of support requirements for component storage, maintenance and retrieval. Work carried out in other European reuse projects (notably REBOOT and DRAGON) will be exploited in this experiment.

2 Reuse Guidelines

To address the problems of creating a base of reusable components which are efficient and reliable without undue cost, we need to involve software project engineers in the reuse process. As we discuss below, we do not believe that this can be accomplished without some organisational costs. However, by providing guidance and information to software engineers about reuse, we believe that we can improve the inherent reusability of software components developed as part of some project.

There have been a number of studies which have been concerned with how to create reusable components [3-5]. These fall into two classes:

1. Language oriented studies which produce guidelines for a specific programming language and suggest how characteristics of that language affect reusability.

2. 'Consciousness raising' studies which try to be language independent. They are of value because they publicise reusability but the guidelines they produce are usually very abstract. For example, all such studies say reusable components should be loosely coupled and highly cohesive. They illustrate these notions with carefully chosen examples. This is absolutely correct. However, the advice is far too general for practical application.

In our experience, the language-oriented studies usually also contain an element of consciousness raising and often do not go into sufficient detail as to the use of the language. Sometimes, they propose very complex approaches to generalisation. Although these may be theoretically interesting, they result in practice in code which is likely to be inefficient and which is hard to understand. Discussions of generalisation using nested Ada generics are particularly incomprehensible to the average software engineer.

These guidelines are interesting to researchers but suffer from practical difficulties when applied in an industrial environment. Such an environment is characterised by tight deadlines, specification-based validation and, in many cases, non-functional time and space constraints on the developed software. All of these mean that it is unrealistic to expect engineers in a normal development process to expend a lot of additional effort to make their software more generic and hence more reusable.

Nevertheless, we believe that it is possible to augment the reusability of a software component to some extent without a great deal of additional effort, and without compromising the non-functional requirements placed on the component. However, if components are to be made truly generic, we believe that the additional effort required must be considered as an organisational overhead.

We have therefore focused on developing two different kinds of reusability guideline:

1. Guidelines which can be applied by software engineers as part of a normal development process. In general, these are intended to increase the flexibility of components and therefore increase the range of applications in which they can be applied.

2. Guidelines which may be applied by 'reuse engineers' whose job is to take existing components, make them more generic and make them available in some organisational component base.

The work we have done has concentrated on developing guidelines for Ada as that is the most widely used development language in the aerospace domain. Similar guidelines could be produced for other development languages. The key criteria which have guided the development of these guidelines are:

1. They must be understandable by software engineers with a reasonable level of Ada programming expertise. There is no point in producing complex guidelines which rely on subtle knowledge of programming language semantics. Very few people understand such guidelines.

2. They must be applicable without a great deal of additional effort. Guidelines will not be applied during development if it means taking longer to develop a component. In essence, they should help engineers make a design choice which has to be made anyway as part of the development process.

3. They must be unambiguous. If guidelines are unambiguous, it is possible to decide whether or not they have been applied without detailed knowledge of the software component. This is important for reuse certification. An assessor of component reusability cannot be expected to have detailed knowledge of all components.

4. They must recognise that embedded systems usually have performance and memory utilisation requirements. Wherever possible, the suggestions made should not degrade the efficiency of a component. If changes are proposed which affect the component's efficiency, this should be explicit so that the efficiency implications may be analysed.

We do not have space here to discuss all of the guidelines which we have proposed. Rather, we describe selected guidelines, their rationale and their practical implications. Our intention is to show that it is possible to produce clear, simple guidelines which, in many cases, lead to efficiency improvements as well as greater potential for reusability. As discussed in Section 3, more complete information on the reuse guidelines is available through the World-Wide Web.

2.1 Improving Component Flexibility

A significant reuse problem is that components are designed for use in a specific environment. Characteristics of that environment are embedded in the component. For example, a component may include absolute values of constants, knowledge about the sizes and organisation of data structures, etc. This is not particularly good programming practice as it reduces the modifiability of the component. It makes reusing the component almost impossible.

Some of our reuse guidelines, therefore, are as much statements of good software engineering practice as anything else. For example, the following guideline concerns the use of arrays in Ada:

> ***Array dimensions***
> *Always use the built-in Ada attributes FIRST, LAST and LENGTH to discover the bounds and size of arrays. Do not pass the array length or bounds as a parameter to a procedure or function . Never make assumptions in a procedure or function about the bounds or the length of the array which is being processed.*

The rationale for avoiding the embedding of array dimensions in a component is, we hope, obvious. Such assumptions integrate the component very tightly with a version of a particular application. The rationale for avoiding passing array dimensions as parameters is perhaps less obvious. Indeed, in a language such as C, a recommendation would probably be that the array dimensions should be passed as parameters.

It is unnecessary to pass array dimension information as parameters in Ada because the language includes built-in attributes to recover this information. Passing it as parameters introduces additional overhead in the procedure call so following this guideline may improve system efficiency. It has also more scope for user error where incorrect values are accidentally introduced. We know that using the built-in facilities will always return the correct information.

A further example of a guideline which improves flexibility is:

> ***Parameter passing***
> *Always define the formal parameters to a procedure or function so that they are initialised by the caller. This means that the parameter passing mode should always be* ***in*** *(for read-only parameters) or* ***in out****. Do not use the* ***out*** *mode.*

If initial values to parameters are assigned in a procedure, these initial values usually reflect the environment for which the procedure was originally defined. We have already discussed the inadvisability of this as far as reuse is concerned. A further reason for avoiding the **out** mode of parameter passing is that its semantics are not well-defined.

2.2 Improving component genericity

A generic component is a component which is specifically designed for use in a variety of different environments rather than in a single environment. There are three dimensions to genericity which must be considered:

1. Adaptability. It should be possible to adapt a generic component without modifying the source code.

2. Understandability. In some cases, it is necessary to modify a component before it can be reused. To make such modifications, it should be possible to understand the operation of the component without knowledge of the application where the component was developed. Of course, knowledge of the domain abstraction modelled by the component may be necessary.

3. Completeness. A generic component should provide complete coverage of the domain abstraction which is modelled by the component.

In this section, we discuss examples of guidelines to improve adaptability and understandability. In Ada, packages are normally used to model domain abstractions. Completeness, therefore, depends on the operations which are provided by a package. This is discussed in the following section.

The generic facilities in Ada have been specifically designed to support adaptability without run-time overhead. We cannot describe Ada generics in detail here but, in summary, the Ada generic facilities allow you to construct versions of packages and subroutines where the sizes of structures and the types of their components may be expressed as parameters. For example, it is possible to define a generic abstract data type as follows:

```
generic
    type Element is private ;
    type Size is (<>) ;
package VECTOR-ADT is
    type VECTOR is array (Size) of Element ;
    ...
```

Generic definitions are instantiated at compile-time rather than at run-time. For example, the above declaration of VECTOR could be instantiated as follows to created a vector of natural numbers:

```
package NAT-VEC is new VECTOR -ADT(Element => NATURAL, Size => 100)
```

In essence, the generic represents a template which is filled in when required. Generics definitions of components can be stored in libraries so represent an excellent foundation for building reusable components. Because they are instantiated at compile-time rather than at run-time, there is no run-time overhead in developing generic rather than specific components.

We have therefore developed a number of guidelines which are intended to help engineers use generics. For example:

Generic data structures
Whenever you define a data structure which is a collection of elements of some kind, always define it as a generic package with the element type as a generic parameter. If the structure is a fixed-length structure (such as an array), define its size as a generic discrete type.

An example of the application of this guideline is the VECTOR-ADT definition given above. The rationale for it is that in many cases, the operations on collections such as arrays, trees, lists, etc. are independent of the type of element in the collection. If we have a generic definition, there is no need for different structures to be defined for each type of element.

An associated guideline which is related to the understandability of a component is concerned with the instantiation of generic elements.

Generic instantiation
Group all instantiations of generics.

This simple guideline means that the reader of a program can easily find the declarations of components which are created from generics. This is helpful when these need to be changed.

As well as supporting the definition of generic abstract data types, Ada also supports the definition of generic procedures and functions. These are subprograms which are partially defined and then instantiated at compile-time. For example, many algorithms which operate on collections such as sorting algorithms, searching algorithms, etc. are independent of the type of the collection. However, these usually rely on generic type functions which allow generic elements to be compared and assigned. This leads to the following guideline.

Generic comparison and assignment
For all generic types used as parameters, always define assignment and equality operators which may also be passed as generic parameters.

This addresses a common reuse problem. Designers of algorithms may incorporate generic elements. However, do not consider that operations such as assignment and equality may not apply when the generic is instantiated to a type which is not an atomic type of the language. Ada allows operator re-definition so this guideline ensures that generic subprograms will work for all types. Reuse engineers may take this responsibility when a component is submitted for reuse.

The generic facilities of Ada may be used, in some cases, to enhance reuse and efficiency. This is illustrated by the following guideline:

Constant subprogram parameters
If a subprogram takes parameters which are constants whose values are always known at compile-time, implement this as an instantiation of a subprogram with generic constant parameters.

Many applications which are designed to operate in different environments. These applications often include procedures and functions which are configured using environment parameters which are set when the system instance for that environment is built. This is good reuse practice. However, it does lead to additional overhead in that long parameter lists may be created.

Ada generics allow this problem to be avoided. By specifying these procedures and functions with references to generic constants rather than constant parameters, the same effect is provided without run-time overhead.

For example, consider the following procedure heading which takes 6 parameters, 4 of which are constant:

```
procedure Set_up ( No_T1_sensors, No_T2_sensors: NATURAL
                   T1_sensor_poll, T2_sensor_poll: Poll_time  ;
                   T1_init, T2_init: Sensor_value) ;
```

This can be rewritten as a generic procedure as follows:

```
generic
    No_T1_sensors, No_T2_sensors: NATURAL ;
    T1_sensor_poll, T2_sensor_poll: Poll_time ;
procedure Set-up (T1_init, T2_init: Sensor_value ) ;
```

The understandability of a component is dependent on many different factors. One critical factor is the use of names in a component. When components are written, they tend to use names which are specific to an application. These names may not be appropriate when a component is reused.

One approach to this problem is for reuse engineers to replace specific names with more general names when a component is made available for reuse. This is the approach used in some Japanese companies who have invested heavily in reuse [6]. Whether or not this is possible, depends on the investment that a company wishes to make. Ada includes a renaming construct which allows entities to be given an alternative name. This means that when a component is reused, the component name may be changed to reflect the particular application where it is used. Component understandability therefore need not be compromised because of reusability requirements. This has led to the following guideline:

> ***Assigning meaningful names***
> *If a component which you wish to reuse does not have a name which is appropriate to your application, use the Ada* ***renames*** *construct to give it an alternative name.*

In order to adopt this guideline, conventions over the use of Ada library packages must be followed. Ada includes a **with** statement to bring library packages into the current scope. Names in these packages are explicitly addressed by preceding them with the package name. However, Ada also includes a **use** statement which brings these names into scope in their own right. They may be referenced without reference to the package name.

This leads to the possibility of name clashes because it may be appropriate to use the same general name in different components. This can cause problems when combined with the above renaming guideline. We therefore recommend that names are always referenced through their enclosing package:

> ***Package usage***
> *Import library units using a* ***with*** *statement. Never use the* ***use*** *statement.*

2.3 Ada package reuse

Ada packages are an encapsulation construct which support information hiding. They can be used to define and implement abstract data types. They are therefore critical for the definition of reusable components.

However, simply defining an abstract data type as an Ada package does not automatically make it reusable. As discussed above, the designer of the package must take care to avoid environmental dependencies in the package. Generics are an important way of factoring out these dependencies and providing package adaptability. A more difficult issue for reuse, however, is concerned with the completeness of the coverage provided by the package operations.

When an abstract data type is implemented as part of a software project, the operations defined on that type are normally those which are required for its implementation environment. However, when the component is reused, it is likely that further operations on that ADT will be required. The coverage offered by the ADT operations reflects the likelihood that the component can be reused without adding additional operations to it.

It is unrealistic to expect developers to implement additional operations on ADTs which they do not need. Furthermore, even if they were to do so, the testing and validation procedures in a project are based on the component specification which may not include the additional operations. Therefore, there is no budget to validate the added operations.

Organisations which are serious about reuse must recognise this problem. They must create a reuse team whose role is to gather packages implementing abstract data types from application projects and modify them to increase their coverage. The ways in which these should be modified is domain-dependent. The coverage required for ADTs representing navigation systems (say) is quite different from that required for engine control systems. Guidelines for the required coverage for each application domain must be developed by domain experts.

As an example of coverage guidelines, we have created guidelines for the domain of abstract data structures. The advantage of working at this level is that these data structures are fundamental and are used in many different kinds of application. A coverage guidelines for abstract data structures describes the classes of operation which should be provided in a generalised component:

Abstract data structures
The operations defined for each abstract data structure should include operations in the following classes:

- *Access operations which are used to inspect the value of structure elements.*
- *Constructor operations to add or remove elements from the structure.*
- *I/O operations to read and write the structure.*
- *Iterator operations to allow each element of the structure to be inspected.*
- *Comparison operations to allow abstract data structures to be compared.*

Of course, the problem with completeness guidelines is that they may result in packages which are large. This may make it impossible to reuse these packages in embedded systems with critical non-functional requirements. One way to address this is to produce generic, complete packages then remove the subprograms which are not required from the source code of the packages. To do so, we have to ensure that the individual subprograms are as independent as possible. This leads to the following guideline:

Subprogram independence
When you implement procedures and functions in a package, try to avoid using other subprograms which are declared in the package interface in that implementation. If it is impossible or not sensible to do so, make sure that you document which interface subprograms are used.

If this guideline is followed, dependencies between operations which are part of the subprogram interface may be avoided. Operations can be removed from the interface and their associated source code deleted from the package body.

3 Information Dissemination

Organisations involved in software development are large, diverse bodies with many sub-organisations within them. They may be geographically dispersed and may use different hardware and software technology for systems development. Information flow between the different parts of the organisation may be imperfect.

In such situations, it is not enough simply to establish a base of reusable components and to expect the information about these components to become widely known in a short time. It is not technically difficult to establish a component database; it is much more difficult to ensure that this database is used and checked for components as part of the development process.

Furthermore, we also must address the problem of how to make the reuse guidelines available to developers. It is all very well to produce a report setting out these guidelines. The reality of practical development, however, is such that we cannot expect software engineers to consult such a report before getting on with their development work. If people have to interrupt their work to look for a report then try to find the relevant parts of the report, they will be reluctant to do so.

We therefore need a dissemination mechanism for component information and reuse guidelines which addresses the following requirements:

1. It must be learnable with minimal effort. Engineers who are involved in software development to tight deadlines simply do not have time to learn complex systems whose value is unknown to them.

2. It must provide interactive access to reuse information across a range of computers. Different groups use different computers such as Unix workstations, PCs and VMS-based systems. It should not be necessary for engineers to learn a different mechanism for each type of machine in an organisation.

3. It must support access to remote information. Because development groups may be geographically distributed, it cannot be guaranteed that their local computers have the required reuse information installed on them.

4. It must be cheap. Organisations will only introduce reuse support (whose benefits cannot be quantified) if involves relatively low start-up costs.

Until relatively recently, it was very difficult to find software which would meet these requirements. However, the explosive growth of the Internet and the development of the World-Wide Web (WWW) has meant that WWW browsers such as NCSA Mosaic are now widely available. Versions of this system are available free or at low-cost for almost all platforms. The user interface is simple and learnable in a few minutes. It is explicitly designed for accessing remote information so long as it is available through standard Internet protocols.

Although intended for general use, it is straightforward to create local or company wide WWW information which is not externally accessible. We therefore believe that this system has the potential to act as a basis for a complete reuse information system. Reuse guidelines can be easily translated to the representation used in the WWW (html) and accessed by engineers from their own machine.

The WWW can also serve as a distributed component repository. There are a number of WWW keyword-based search programs which can search across sites. These can be activated from Web browsers. A component retrieval system based on these keyword searches can therefore be implemented and integrated with other reuse information. Files on remote machines holding the source or object code of components may be collected using file transfer systems such as gopher and ftp. Rather than build a special purpose database for software components, they can be held on the most appropriate system and indexed using WWW facilities. When a component is required, a search can be made and components retrieved.

While we have not yet developed company reuse information systems, based on Web browsers, we are currently creating an APPRAISAL project information base on the WWW[1]. This is providing experience of the advantages and disadvantages of using this system to store reuse information. It also serves as a means of disseminating reuse knowledge to the community at large.

Of course, issues such as security, integrity, and the management of change must be addressed in the development of organisational reuse systems. These are neither simple nor are they exclusively problems of managing information about reuse. We must admit that they are not well-supported by current Web browsers. Nevertheless, we believe that the advantages of using the WWW browsers as a means of accessing and disseminating information are overwhelming. This

[1] The URL for reuse information from the APPRAISAL project is http://www.comp.lancs.ac.uk/computing/research/cseg/projects/APPRAISAL/

approach offers a very cost-effective means of making reusability information widely available within and outside an organisation.

4 Conclusions

The approach which we have described here may appear simple, even simplistic. It does not require any new tools nor does it propose new methods of software development. We readily admit that it has technical limitations. Nevertheless, we believe that our proposals are realistic as far as industrial take-up is concerned. Industry can adopt the reuse guidelines and the dissemination technology without significant capital investment, without changing existing methods and without requiring its software engineers to develop new skills. Imposing new methods and new technology is rarely successful; our approach provides an evolutionary rather than a revolutionary approach to reuse.

We note, however, other changes in development practice which have a significant impact on reuse and which are not addressed by our guidelines. In particular, the use of design methods such as HOOD [7] and associated CASE tools is changing the way in which software is developed. The supporting tools often include a code generator which automatically generates some or all of the code implementing a component from a design specification. The engineer does not have the opportunity of applying the reuse guidelines at the code level. Rather than reuse code components, we must consider the reuse of design elements. The challenge we face, therefore, is to develop further guidelines which apply to designs rather than code.

5 Acknowledgements

Dr M. Ramachandran was closely involved in developing the guidelines for Ada reuse. Dr E. El-galal assisted with the construction of the WWW information system. Support for this project has been provided by the European Commission under the ESSI Initiative (Project ESSI 10452, APPRAISAL).

References

1. McIlroy, M.D. "Mass-produced software components". in *NATO Conf. on Software Eng.* 1968. Garmisch, Germany. 1968.

2. Checkland, P. and J. Scholes, *Soft Systems Methodology in Action*. Chichester: John Wiley & Sons. 1990.

3. Braun, C.L. and J.B. Goodenough, *Ada Reusability Guidelines*, Softech: 3285-2-208/2. 1985.

4. Booch, G., *Software Components with Ada: Structures Tools and Subsystems*. Menlo Park, Ca.: Benjamin Cummings. 1987.

5. Gautier, R.J. and P.J.L. Wallis, ed. *Software Reuse with Ada*. Stevenage, UK: Peter Perigrinus. 1990.

6. Matsumoto, Y., "Some Experience in Promoting Reusable Software: Presentation in Higher Abstract Levels". *IEEE. Trans. on Software Engineering*, **SE-10**(5), 502-12. 1984.

7. Robinson, P.J., *Hierarchical Object-Oriented Design*. Englewood Cliffs, N.J.: Prentice-Hall. 1992.

Process Quality Problems from the Point of View of the User and of the Purchaser of Software

Antonio Cicu

METRIQS S.r.l, QualityLab Consortium, Italy
Milano, Via don Gnocchi, 33, phone +39-2-48701795

Abstract. The user/purchaser of software products makes investments of resources in software as one of the means for improving its own products and business.The level of improvement obtained depends, among other factors, on the quality level of the acquired software. Making reference to the relevant standard, normative and methodological documents, the paper illustrates how the quality level of the subcontracted software depends not only on the quality of the software supplier's process, but also on the quality of purchaser/user's own production/business processes. Being the purchaser/user the primary responsible of the preparation of the software requirements, which specify why the purchaser wants to invest in software, and what the software must do to improve the purchaser's products or purchaser's business process, the paper highlights the quality management activities which have to be performed by the purchaser, or monitored on the side of the supplier, in order to ensure a successful achievement of the purchaser's business goals.

1 Introduction

In a context where an organization acquires a software product for using it for its own purposes, the paper presents *the point of view of the purchaser/user* with regard to the problems of the quality of processes which influence the quality of the acquired product.

The following aspects of process quality, all relevant from the user/purchaser's point of view, are illustrated:

1. *purchaser's processes* have to be periodically assessed to determine the why, what, where and how for the software to be installed and integrated into the purchaser's products or business process, and to define the advantages in terms of costs/benefits
2. *how software requirements* have to be *expressed* in terms related to purchaser's product needs or to expected purchaser's process improvements
3. how the *software supplier's process* has to be monitored, in order to ensure that the acquired software meets the software requirements
4. the purchaser's role in the *contractual and cooperative relations* with the supplier.

For reasons of text brevity, the following acronyms are used throughout the paper, also in composite terms: "SW" in place of "SOFTWARE" or "software"; "CUS" in place of "CUSTOMER" or "customer", but meaning also "user", "end-user", "purchaser"; "SW_P/U" in place of "SOFTWARE_PURCHASER/USER"; "SW_Reqs" in place of "software requirements"; "SW_SUP" in place of "SW_SUPPLIER".

2 The "SW_SUPPLIER to SW_PURCHASER/USER" Supply Chain: Global View

Figure 1 provides a *global view of the commercial context* to be considered when examining the process quality problems which are relevant to the purchaser/user of a software product.

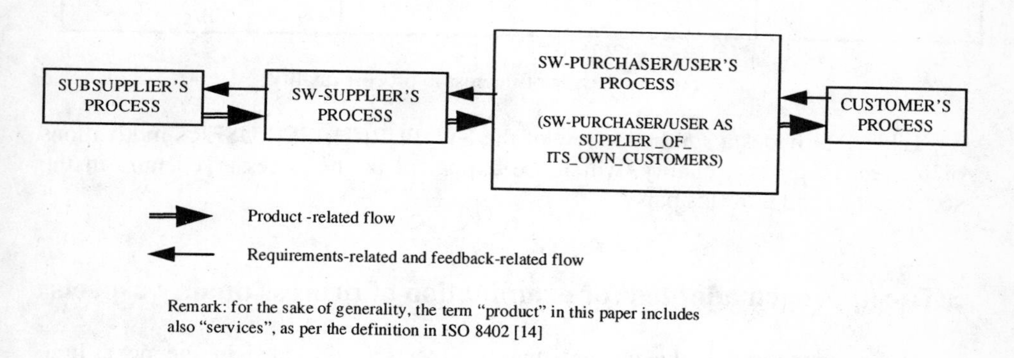

Fig. 1. The SW_SUPPLIER to SW_PURCHASER/USER supply chain

The content of Figure 1 is conceptually equivalent to the "supply chain" schemas provided by the standard ISO 9000-1 [15] (see its Table 1 and Fig. 2).

The specific aim of Fig.1 is to point out that a discussion on the interests in process quality by the SW_PURCHASER/USER requires that *attention be given* to its organization not only as a CUSTOMER of SW_SUPPLIER, but also and *first of all to its role of SUPPLIER_to_its_own_CUSTOMERS (alias: SUPPLIER_to_CUS).*

Levels of interest to process quality by the SW_PURCHASER/USER point of view		
Level of interest	Process considered	Comments
Very high, primary, strategic.	The process of SW_PURCHASER/USER's organization, as SUPPLIER_of_its_CUSTOMERS	Quality of this process is a prerequisite for producing *software requirements which fit the company's goals*.
Very high, but secondary and instrumental to the above.	The process of SW_SUPPLIER's organization.	Quality of this process is a prerequisite for producing and servicing *software which fits the above requirements*.

Tab. 1 - Levels of interest to process quality.

Table 1 provides a synthetic view of the SW_PURCHASER/USER's motivations of interest to process quality, which are explained in the necessary details in the sections 4,5, and 6 of this paper.

3 The approach adopted for examination of process quality aspects

The examination of the process quality aspects is provided in the next three sections of the paper, corresponding to the following *three periods of the software life-cycle*:

- Planning and Definition phases
- Design, Implementation, Validation, Delivery, Installation phases
- Operation and Maintenance phases.

The examination is conducted commenting three tables (Tables 2, 3, 4), where, per each of the mentioned life-cycle periods, a *synthesis* is provided of the *process quality aspects which are of interest for the SW_P/U*.

The tables are divided in two parts: the first part gives a view of process quality aspects as they are dictated by relevant standards and directives, the second part provides references to consolidated (or under active development) methods which can help in implementing or improving the process capabilities required by standards and directives.

3.1 Relevant standards and directives providing guidance for quality management, assurance, monitoring in a given period.

This part of the tables is divided in two columns, showing which are the main goals for which ISO 9000 standards and AIPA directive [2] dictate obligations or provide guidance.

3.1.1 ISO 9000 Quality management standards

The documents ISO 9004-1 [18] and ISO 9004-4 [19] are the references for the SW_PURCHASER/USER's organization, in the implementation of its own internal Quality management system.

The documents ISO 9001 [17] and ISO 9000-3 [16] are the references for the SW_SUPPLIER's organization, because of the contractual situation characterizing its relation with SW_P/U.

3.1.2 AIPA directive

This directive [2], issued by the Italian National Authority for high-relevance IT projects in Public Administration, is a reference example of the type of self-regulation policies that can be adopted by the management of a given community of users. The directive provides a synthesis of good quality management criteria to be adopted in the projects, and takes explicitly inspiration from ISO standards (referencing ISO 9000 certification and ISO 9126 [20]). Through the execution of the monitoring activities, dictated by the directive for high-relevance projects, the Public Administration wants to ensure that public money investments in software produce real added values for the Public Administration services.

3.2 Helps for Implementation of Process/Product quality measurement, process assessment, process improvement

This part of the tables is divided in three columns (dedicated to SEI's CMM, SPICE project, other supports) showing which are the sources from where the organization can get helps in implementing or improving its own processes for achieving the goals shown in the left part of each table.

3.2.1 SEI's Capability Maturity Model

The Software Engineering Institute's Capability Maturity Model (SEI's CMM, or CMM) ([24], [25]), has been used for reference because of the richness in details of the model, and because being this model in substantial way the reference origin of other methods developed for process assessment and improvement.

The tables provide references to the specific Key Process Areas [25] which specify details useful in approaching the process aspects covered by each table.

3.2.2 SPICE project

The SPICE project ([29], [31]) has been referenced in the tables, again for the richness of its process model, but also because it constitutes a relevant effort by ISO for the standardization of process assessment and improvement requirements. The most expert authors of the major available methods are contributing to the project: this is a condition which ensures very good probabilities that the coming standard (planned for 1996) will be really a merge of the best practices in this area. Currently the project has undertaken the first trial phase, for a first evaluation of its Baseline Practices Guide [30]. The tables of this paper provide references to the relevant Processes described in the Baseline Practices Guide. See [31] for an overview of the project.

3.2.3 Other supports

This column of the tables gives references to various kinds of methods or standards, considered particularly useful for an effective implementation of quality management, in the cases that, in the opinion of the author, such methods or standards provide additional help versus what suggested by CMM and SPICE. Specific comments for each referenced item are provided in Section 4,5,6. In this column the reference "Other Process Assessment and Improvement Methods" regards the following methods: BOOTSTRAP, Software Technology Diagnostic (STD), HealthCheck/SAM, Trillium, Software Quality and Productivity Analysis (SQPA). These methods are mentioned in [29] as the major methods which, together with CMM, have contributed to SPICE, and are recalled here for completeness. Detailed references to methods' documents are in [30]; an analysis of these methods is beyond the scope of this paper.

4 The SW_PURCHASER/USER as SUPPLIER_to_CUS with its own Industrial Process (Planning and Definition Phases)

The objectives of this period, for the SW_PURCHASER/USER, are:
- production of the software requirements
- production and assignment of the contract.

The process aspects considered cover what is necessary to ensure the quality of these two deliverables.

In general the SW_P/U's company is not a software producer, but a business actor operating in market sectors (hardware, processed materials, services) different from software.

This paper considers explicitly the case of a SW_P/U which does not possess competence in Software Engineering neither in related technologies, and illustrates which are the aspects of quality management culture which have to be acquired and mastered by SW_P/U as user and purchaser of software products and systems. This point is strategic for every kind of productive or servicing community, being software a product which is pervading more and more all economic sectors. In fact there are typical software engineering processes that must be mastered also on the side of any SW_P/U (e.g.: initial planning and definition processes, contract management, acceptance, service, costs/benefits evaluation, etc.).

There are two main categories of scenarios (see Fig. 2) where the SUPPLIER_of_its_own_CUSTOMERS decides to purchase a software product:

- 1-st scenario: software is used and integrated as component embedded into a final product or system
- 2-nd scenario: software is used for internal support purposes of productivity/quality improvements or resources saving, required for achieving the selected company's goals.

In both cases the SW_P/U faces the problem of how to ensure that its software investment decision is well oriented to, and driven by, its business goals.

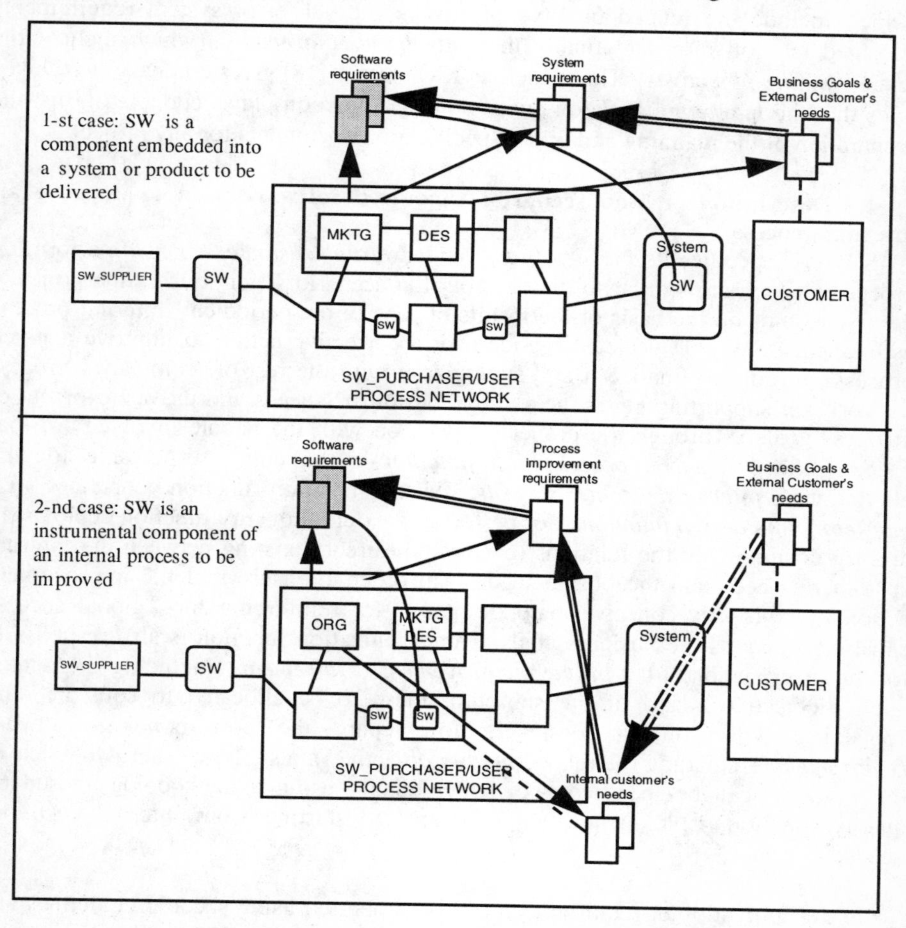

Fig.2 Two possible application scenarios for the SW_PURCHASER/USER

Considering the indications of Table 2 (1-st column), there are some differences in the ways that SW_P/U follows in producing the software requirements in the two scenarios.

4.1 Description of 1-st scenario: acquired software embedded into a final system or product

SW_P/U specifies its *company's goals* (i.e. *business goals*, and *external customers' needs)*: a good execution of this basic activity implies that SW_P/U has already in place a working channel of communication with its customers, to collect their feedbacks and requests, and assesses periodically the *customers' satisfaction* (see Section 6).

Once the periodical preparation or updating of the *strategic plan* is done for new products and/or new services, the SW_P/U specifies, possibly using a methodology based on quality function deployment (QFD [1], [32]), the *system requirements* (which include the related quality objectives), the subset of system requirements allocated to software, and finally the *software requirements* (which include the related *quality objectives*). The guideline ISO 9004-1 [18] gives emphasys to the key roles that the *marketing* function and the *design* function play, cooperating in the preparation of the planning and initial specification documents for any project.

4.2 Description of 2-nd scenario: acquired software used to improve an internal process

SW_P/U specifies its *business goals* and *external customer's needs* performing processes which are similar to the ones of 1-st scenario. Then, during the project's life-cycle, and on the basis of the results of one of the periodical internal process assessments, SW_P/U decides to undertake a corrective action to improve a given process, in order to finalize it and synergize it with the rest of company's process network, for supporting better the external customer's needs and the achievement of business goals. Through the proper cooperation with the people involved in that process, the company's *organization function* is the entity responsible for the elicitation of *internal customer's needs*. The organization function cooperates with *marketing and design functions* for performing external quality function deployment and for documenting the relations (even if not direct) existing between the internal customer's needs and the business goals, in order to ensure that the improvement action is expressely conceived and designed for reinforcing the global success capability towards the business goals. The organization function is also responsible for the specification of the *process improvement requirements* (including the related quality objectives), and of the subset of these to be allocated to software, and cooperates with software designers to prepare the corresponding *software requirements* (including the related *quality objectives*); also these other two levels of requirements can be specified with better quality using a methodology based on quality function deployment ([1][32]), applied starting from internal customer needs.

The specification of requirements can be supported using specific guidelines and templates (like the ones available in IEEE [10] and ESA [6,7,8] standards), and applying methodologies of requirements elicitation [27] to minimize the risk of not considering important items. The specification of quality goals and subgoals has to be joint to the specification of the evaluation criteria and metrics, and can be supported by methodologies like quality function deployment, ami [3,4], Goal/Question/Metrics (GQM,[5]); ami supports also the specification of related metrics.

Requirement specification activity is concluded defining for each requirement a *priority level* value, performing *feasibility and profitability analyses*, and evaluating and measuring requirements quality attributes, using the attribute tree shown in Fig.3 as guidance.

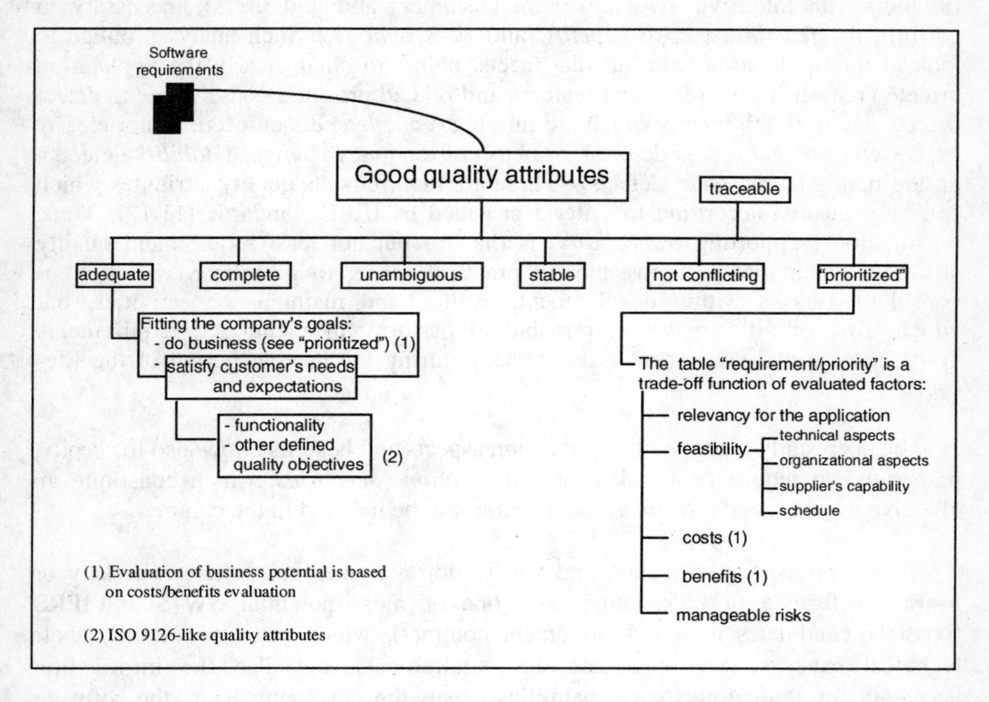

Fig. 3. The attributes of software requirements, necessary to make a good quality design input

ISO standards, aiming to promote a nonconformities prevention project practice, impose explicitly that requirements be reviewed, before design, to verify if they are *adequate*, *complete*, *unambiguous, unconflicting*; the same standards demand also implicitly that the requirements be verified for *feasibility* and *profitability*. *Adequacy* means capability of fulfilling the company's goals: the capability of supporting business goals is verified via *costs/benefits analyses* (for which guides are available in documents [21], [23], and some indications in ISO 9004-1 [18]); the capability of supporting the customer's needs is verified by means of evaluation of technical factors, like the ones provided by the ISO 9126 model [20]. Before approaching the evaluation of quality attributes like *completeness, unambiguousness, unconflictingness (consistency), stability* it is necessary that the definition of *requirements priorities* is performed. The priority level of a requirement is the value of a management trade-off function which takes into account more factors: first of all the relevancy of the requirement for the application, then the feasibility, the costs, the benefits, and finally the evaluation of associated *risks*. Risks evaluation and management can be helped by some available documents [22,28].

Once the different levels of requirements have been elicited and specified, what ensures and consolidates the global requirements quality is the set of activities (including an intensive involvement of customers and end users) necessary to perform the *feasibility, costs/benefits*, and *risks analyses*. Such analyses oblige to look at the application from all due facets, oblige to elicit clearly the application oriented reasons of the requested features and related pros and cons, allow to detect defects due to requirements which are missing (*completeness* defects), or not clearly expressed (*ambiguousness* defects), or object of too many changes (*stability* defects), or conflicting (*consistency* defects). These are requirements quality attributes which can be measured according to criteria provided by IEEE standards [11,12]. Trace information (supporting *traceability*) is the last but not least requirement quality attribute which makes it possible, in practical sense, to perform verification or consultation tasks within development, testing, and maintenance activities, but which, first of all, makes it possible to perform all mentioned requirements verification and measurement tasks, already during the current period of the life-cycle.

The deep study made of the application aspects and benefits allow also to specify, in parallel to requirements, also the *application scenarios*, which constitute an effective way to specify the *acceptance criteria* to be inserted in the contract.

The mentioned specification and verification activities are, where necessary or useful, performed in cooperation with one or more potential SW_SUPPLIERS (possibly candidates to the development contract), which provide not only their technical expertise to consolidate the requirements, but also the information necessary for evaluating their capabilities versus the development of the software fitting the requirements. Guides for capability evaluation are available, based on the CMM model [26], and one more guide is under development by SPICE project. The SW_P/U is now ready for selecting a SW_SUPPLIER and for assigning a *contract*, in order to develop a specified, selected subset of the verified and prioritized software requirements.

Table 2 - Process Quality Aspects in SOFTWARE_PURCHASER/USER's organization (Planning and Definition Phases)

Relevant standards and directives providing guidance for quality management and monitoring		Helps for Implementation of Process/Product quality measurement, process assessment, process improvement		
ISO 9004-1, 9004-4	AIPA	SEI's CMM	SPICE	Other supports
• Define *company's goals*: satisfy customer's needs, achieve business objectives • Derive *strategic plans* for new products, new services, and/or internal process improvement • Define *system requirements*, and *which of them are allocated to software*. Define the corresponding adequate *software product requirements, or the software-based process requirements*, including for all above the *quality objectives* and the related evaluation criteria and metrics • Perform and document *feasibility and profitability* (covering *costs, benefits, risks*) *analyses*, supporting them (eventually in cooperation with candidate software suppliers) the necessary and sufficient software specification and design activities • *Evaluate select the software supplier(s)* • Develop/commit a *contractual agreement* with the selected software supplier(s), emphasizing prevention of nonconformities • Perform *continuous quality improvement*	• Strategic Planning • Feasibility Study and Supplier's evaluation • Contract Agreement & Contract Assignment • Authorization to project execution and monitoring execution	• Requirements management • Software project planning • Software subcontract management • Training program • Integrated software management • Software product engineering • Intergroup coordination • Software quality management • Peer review • Defect prevention • Technology change management	• CUS.1 Acquire Software Product and/or Service • CUS.3 Identify Customer Needs • ENG.1 Develop System Requirements and Design • ENG.2 Develop Software Requirements • PRO.6 Manage Risks • ORG.2 Define the Process • ORG.3 Improve the Process • CUS.2 Establish Contract	• SEI reports on Requirements elicitation • IEEE, ESA standards for requirements specifications • SEI reports on Risk management • QFD • ami • ISO 9126 standard • Cost/benefits analysis guides • Other Process Assessment and Improvement methods

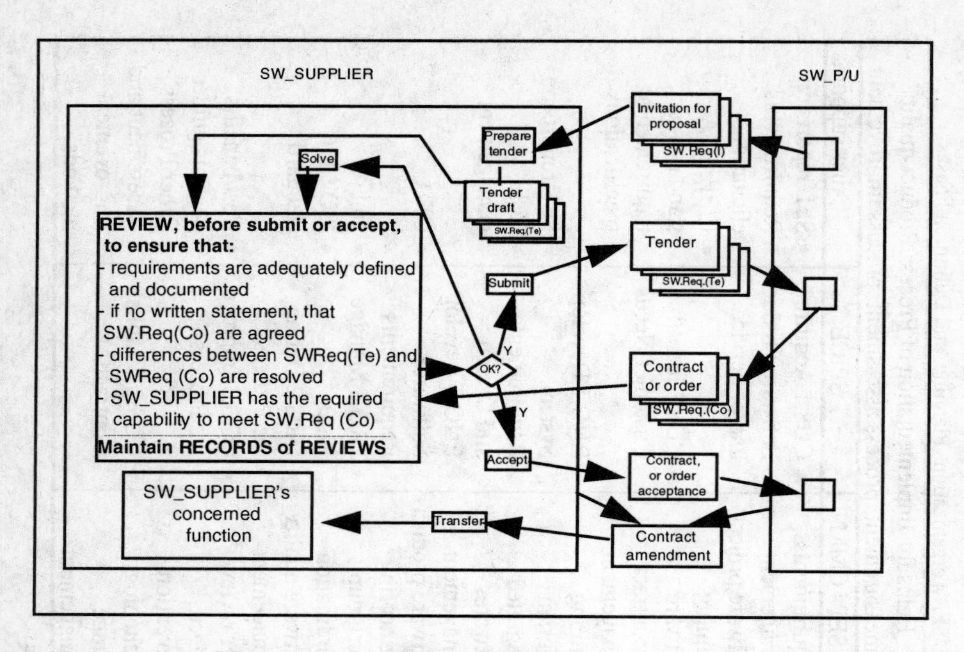

Fig.4. ISO 9001. The requested flow for contract preparation, review, amendment: the obligations on supplier's side.

Fig. 4 illustrates the proactive process through which the SW_SUPPLIER ensures the SW_P/U that a nonconformities prevention approach is applied, as early as at time of tender submission. Fig. 4 gives evidence of ISO 9001 stated requirements on the review obligations on the side of SW_SUP, before submission of the tender and before the acceptance of the contract or order, and on support obligation for contract amendments.

In addition to *acceptance criteria*, the contract is the place where to formalize agreements on other process quality aspects, important for the SW_P/U for managing the relations with SW_SUP: *joint reviews* plan, *quality assurance*, *verification methods*, *settlement of disputes*, the *channel of communications* with SW_SUP.

5 Process Quality Aspects in SOFTWARE_SUPPLIER's Organization (Design, Implementation, Validation, Delivery and Installation phases)

The objectives of this period are to build a software product which fits the requirements, and to build the confidence on product quality by the SW_P/U. The basic means are to perform engineering and monitoring activities which conform to the standards, directives, and process models referenced in Table 3. Details for the application of the implementation guidelines and for monitoring, regarding the supplier's *quality system*, the *project development plan*, and the project *quality plan*, are available in the referenced documents [2, 9, 16, 17, 18, 25, 30].

Main purpose of this section is to give *emphasys to the following process aspects,* which are more important than others for the SW_P/U, because they regard topics which are *directly useful to*, and *easily verifiable by, the SW_P/U*:

- as early as possible define with full detail the software *user interface*, and develop *prototypes* which on one side allow to verify the interface with the customer in a live way, on the other side allow to anticipate the development of initial versions of system and acceptance tests
- use the *application scenarios*, used already as specification of acceptance tests, *in support of review of architecture completeness and consistency*, describing in proper tables the dynamic internal behaviour of the architecture versus the user visible operations of the scenarios
- perform *joint reviews* as agreed in the contract, to monitor *progress* and especially *risks assessments* and *quality measures*
- monitor the *test coverage* of the different levels of tests, especially for system and acceptance tests (templates for test documentation, usable in support of this activity, are available in IEEE standards [9])
- perform all the activities using modern methods and technologies, but first of all document carefully the *tracing of requirements* to every level of software work products
- use defect detection and removal measurements, and Quality Records, to verify that the *support capability* necessary for the Operation and Maintenance phases was built timely by SW_SUP, and practiced internally during the current period
- monitor SW_SUP's efficiency in supporting *requirements changes* and *contract amendments*
- perform *acceptance* activity as agreed in the contract

Table 3 - Process Quality Aspects in SOFTWARE_SUPPLIER's Organization (Design, Implementation, Validation, Delivery and Installation phases), of relevant interest for the SOFTWARE_PURCHASER/USER.

Relevant standards/directives providing guidance for external quality assurance and monitoring in contractual situations		Helps for Implementation of Process/Product quality measurement, process assessment, process improvement		
ISO 9001, 9000-3	AIPA	SEI's CMM	SPICE	Other supports
• Demonstrate that the supplier's *quality system*, properly documented in a quality manual, is conformant to ISO 9000 requirements. • Prepare a *project development plan* documenting the necessary development, verification, validation, packaging and delivery activities required to produce a software product fitting the contractual requirements. The plan describes also the risks associated to the projects, and the related contingency measures • Prepare a *project quality plan*, documenting all the quality verification related activities, based on defect prevention oriented criteria and approach, specifying the selected quality objectives and measurement criteria • Conduct *joint reviews* • Perform and document the planned *quality measurements* • Perform *acceptance tests* • Perform *continuous quality improvement*	• Monitoring on Supplier's Process • Monitoring on Project's Management • Monitoring on Product's Quality	• Software project planning • Software project tracking and oversight • Software subcontract management • Software quality assurance • Software product engineering • Software quality management • Peer review • Defect prevention • Quantitative process management	• PRO.2 Establish Project Plan • CUS.4 Perform Joint Audits and Reviews • PRO.4 Manage Requirements • PRO.5 Manage Quality • PRO.6 Manage Risks • SUP.1 Develop Documentation • ENG.6 Integrate and Test System • CUS.5 Package, Deliver, and Install the Software • ORG.2, ORG.3 as in Table 2	• IEEE, ESA standards for test specifications, for quality measuresments • SEI reports on Risk management • QFD • ami • ISO 9126 standard • Other Process Assessment and Improvement methods

6 Process Quality Aspects in the Operation and Maintenance Phases

The objectives of this period are to support operation and maintenance of software, to verify the actual investment added value, to assess customer's satisfaction. The documents referenced in Table 4 provide detailed requirements and process models for the implementation or improvement of the required processes [2, 16, 17, 18, 25, 30]. *Field quality measurements* are a key activity in this period, mainly for software anomalies analyses and defect prevention analyses. IEEE standards provide a help for these analyses [13].

The *investment in software must be sustained* by SW_P/U management, supporting the (internal or external) customers with adequate *training* program on the application objectives and features. The application scenarios, developed as acceptance tests, can be used also for training of the SW_P/U personnel on the novelties of the application or system. Only adequately educated and motivated end users can exploit fully the product features. Care must be applied to support the introduction of the new product as planned, and to sustain the initial process improvement or product innovation objectives.

Particularly important is the *verification of the actual investment added value*. The evaluation model is the same as the one used for the costs/ benefits estimation during the Planning and Definition phases. The same guides [21, 23] can be used for help.

Where the actual added value evaluation differs from the expected one, a diagnose of possible causes has to be made.

Customer's satisfaction has to be periodically assessed to know timely possible causes of loss of confidence by customers, and to determine which corrective actions have to be decided to consolidate customer's confidence, to keep current business and initiate a new one. A major issue for SW_P/U is to follow the evolution of interests in the customer's company, in order to understand how to riorient its own product features. The SW_P/U understands to have the necessity to know as deeply as possible the *customer's process quality problems*, because they will be the source of new needs, or of complaints on the present products. A good mastering of customer's process problems (as well as of its market and its competitors) allows SW_P/U to achieve more results:

- understand better and timely the customer's needs
- better ability to show to customers how SW_P/U's products fit the customer's needs
- to plan in time the evolution of SW_P/U' products and show to customers that those products do have a future.

Table 4 - Process Quality Aspects in the Operation and Maintenance Phases, of relevant interest for the SOFTWARE_PURCHASER/USER.

Relevant standards and norms providing guidance for quality management, assurance, monitoring in contractual situations		Helps for Implementation of Process/Product quality measurement, process assessment, process improvement		
ISO 9001, 9000-3, 9004-1, 9004-4	AIPA	SEI's CMM	SPICE	Other supports
• Support operation and maintenance of software • *Training* the *customers and end users on the applicationnew* features and related objectives and benefits • Provide *customer service* and perform *problem resolution* • Verify the *actual investment added value* • Assess *customer satisfaction* • Perform *field quality measurements* • Perform *continuous quality improvement*	• Monitoring on Investment value (Cost/Benefits analysis)	• Requirements management • Software quality assurance • Software configuration management • Training program • Software product engineering • Software quality management • Defect prevention • Technology change management	• CUS.6 Support Operation of Software • CUS.7 Provide Customer Service • SUP.4 Perform Problem Resolution • ENG.7 Maintain System and Software • CUS.8 Assess Customer Satisfaction • ORG.2, ORG.3 as in Table 2	• IEEE standards for software anomalies classification • ami • ISO 9126 standard • Cost/benefits analysis guides • Other Process Assessment and Improvement methods

7 Conclusions

Process quality expertise is a set of competences and skills which support a company in setting up and improving its own process network for a higher product quality and a more successful business. The paper has shown why and how a company, which needs to acquire software in support of its business, has to master also the process quality issues of its customers as well as of its software suppliers. The paper has illustrated the practical advantages of expressing the requirements, from customer's needs to software requirements, in a terminology centered on, and traceable to, its own business goals and customer's application needs.

References

1. AICQ Italia CentroNord - A. Salvini, docente - Text in support of the course on "Process Control" (Milano, October 18-19, 94)
2. AIPA, Autorità per l'Informatica nella Pubblica Amministrazione - Circolare 5 agosto 1994, n.AIPA/CR/5 - Gazzetta Ufficiale della Repubblica Italiana, n.191, 17-8-1994
3. ami (application of metrics in industry), Handbook - A quantitative approach to software development - The ami consortium, c/o The ami User Group, CSSE, South Bank University, 103 Borough Road, London SE1 OAA
4. A. Combelles - Quantitative Approach to Software Process Improvement - Springer Verlag, Proceedings of Objective Quality 1995, The 2nd Symposium on Software Quality Techniques and Acquisition Criteria, Florence, Italy, May 29-31, 1995
5. Basili, V. and Weiss, D. M. - A methodology for Collecting Valid Software Engineering Data - IEEE Transactions on Software Engineering, 1984
6. ESA PSS-05-0 Issue 2, February 1991 - European Space Agengy Procedures, Specifications, and Standards for Software Enginering, Issue 2
7. ESA PSS-05-02 - European Space Agengy, Guide to the User Requirements Definition Phase
8. ESA PSS-05-03 Issue 1, October 1991 - European Space Agengy, Guide to the Software Requirements Definition Phase
9. IEEE Std 829-1983, Standard for Software Test Documentation
10. IEEE Std 830-1993, Recommended Practice for Software Requirements Specifications
11. IEEE Std 982.1-1988, Standard Dictionary of Measures to Produce Reliable Software
12. IEEE Std 982.2-1988, Guide for the Use of Standard Dictionary of Measures to Produce Reliable Software
13. IEEE Std 1044-1993, Standard Classification for Software Anomalies

14. ISO 8402:1994 Quality management and quality assurance - Vocabulary
15. ISO 9000-1:1994 Quality management and quality assurance standards - Part 1: Guidelines for selection and use.
16. ISO 9000-3:1991 Quality management and quality assurance standards - Part 3: Guidelines for the application of ISO 9001 to the development, supply and maintenance of software
17. ISO 9001:1994 Quality systems - Model for quality assurance in design, development, production, installation and servicing
18. ISO 9004-1:1994 Quality management and quality system elements - Part 1: Guidelines
19. ISO 9004-4:1994 Quality management and quality system elements - Part 4: Guidelines for quality improvement
20. ISO/IEC 9126:1991 - Information Technology - Software Product Evaluation - Quality characteristics and guidelines for their use
21. NIST (National Institute of Standards and Technology) - FIPS PUB 64, Guidelines for Documentation of Computer Programs and Automated Data Systems for the Initiation Phase, 1979
22. NIST (National Institute of Standards and Technology) - FIPS PUB 65, Guideline for Automatic Data Processing Risk Analysis, 1979
23. Nocentini, S. Il sistema di qualità del software. Il processo di pianificazione, realizzazione e controllo. - ETASLIBRI, Collana di Informatica, 1993
24. Paulk, M. C. and others - SEI-93-TR-024 - Capability Maturity Model for Software, Version 1.1 - Software Engineering Institute, Pittsburgh, PA
25. Paulk, M. C. and others - SEI-93-TR-025 - Key Practices of the Capability Maturity Model, Version 1.1 - Software Engineering Institute, Pittsburgh, PA
26. SCE Staff - SEI-94-TR-006 - Software Capability Evaluation, Version 2.0, Method Description - Software Engineering Institute, Pittsburgh, PA
27. Kang, K. C. and Christel M. G. - SEI-92-TR-012 - Issues in Requirements Elicitation - Software Engineering Institute, Pittsburgh, PA
28. Higuera, R. P. and others - SEI-94-SR-001 - An introduction to Team Risk Management - Software Engineering Institute, Pittsburgh, PA
29. SPICE project (Software Process Improvement &Capability dEtermination) - Tutorial, at Fourth European Conference on Software Quality, October 17-20, 94, Basel, Switzerland
30. SPICE project - Baseline Practices Guide - Version 1.00, September 1, 94 - Doc Ref BPG\TP\BPG - (Document in internal use in SPICE Project)
31. T. Coletta -The SPICE Project: An ISO Standard for Software Process Assessment, Improvement and Capability Determination - Springer Verlag, Proceedings of Objective Quality 1995, The 2nd Symposium on Software Quality Techniques and Acquisition Criteria, Florence, Italy, May 29-31, 1995
32. Zultner, R. E. - Software Quality Function Deployment: Why, What, and How - Tutorial, at Fourth European Conference on Software Quality, October 17-20, 94, Basel, Switzerland

Software Quality of Use: Evaluation by MUSiC

Laura Binucci
Data Management S.p.A.
Centro Direzionale Colleoni - Via Paracelso, 2
20041 Agrate Brianza, MILANO - ITALY
Phone: +39 39 6052.1 - Fax +39 39 6057497

Abstract. Usability is becoming more and more a fundamental aspect of the software quality. This is due to the fact that computers are now used by people with a broad range of knowledge and experience and for a large variety of tasks. Furthermore, since nowadays technology has advanced to such an extent that every software industry is able to provide more sophisticated features, the quality of use becomes the real competitive factor that differentiate products in the software market. Considering that the concept of usability involves not only software, machines and documentation but, above all, people who interact with them, it remains a concept difficult to evaluate, to measure and to manage by the highly skilled technical people of the software industries. The MUSiC method provides the means to specify usability requirements and to measure how well those requirements are met. The MUSiC method can be applied at different stages of the software lifecycle, after a training course, without specific Human Factors or Human Computer Interaction experience and in a cost effective way.

1 Introduction

During the past twenty years the diffusion of the personal computers and falling hardware prices have caused broad groups of people to come into contact with computers [12]. Computers are now used by people with the more disparate range of knowledge and experience to perform a large variety of tasks. Furthermore technology has advanced to such an extent that the software products provide more and more sophisticated features.

For these reasons user interfaces have become a strategic part of computers. The way users interact with computers determines the efficiency, the effectiveness and the satisfaction with which they perform their tasks using computers.

It is therefore essential computer software to be easy to understand, to learn and to use.

The benefits deriving from an appropriate designed computer system are both for users and developers [2]. In fact from the users' point of view usable products increase productivity and satisfaction and reduces costs. On the other side, in the actual software market, the difference among the features the software industries are able to provide are becoming smaller and smaller and therefore software suppliers can differentiate their products just through of their quality of use characteristics. Furthermore the recent political initiatives are clearly taking into consideration software usability as it appears from the International Standards (ISO 9126 *IT - Software product evaluation - Quality characteristics and guidelines for their use* [10], ISO CD 9241 - Part 11 *Ergonomics requirements for office work with visual display terminals - Guidance on Usability* [9]) and from the Council Directive of May 29, 1990 on the minimum safety and health requirements for work with display screen equipment (90/270/ECC) that states that:

- Software must be suitable for the task
- Software must be easy to use
- The principles of software ergonomics must be applied

The conformance to these International Standards and requirements of the European Directive is going to constitute the prerequisite to participate to public procurements [2].

Unfortunately the concept of usability can not be limited to the software technical aspects. It implies the users' involvement and therefore it has to be related not only to the software quality characteristics but even, and above all, to the software quality perceived and performed by the humans who interact with the software in a specific environment to achieve specific goals. This is particularly hard if we consider the highly technical skill of people in the software industries not used to deal with Human Factors and Human Computer Interaction thematics. One of the most difficult problems that implies products not to be usable is that "developers are not users and users are not developers" [11]. This means that it is very difficult for developers to predict the users' behaviour because they are more technically skilled than users could ever be. On the other side users have a different approach to the software which is the one that the tasks, the objectives and the environment of their work and their feeling impose to them.

To overcome the difficulties due to the impossibility to limit usability to a pure software technical aspect and to the existing gap between users and developers, it is necessary, first of all, to sensitize developers to the problematics related to usability and then to provide them with the means to identify in a clear way the requirements a product has to satisfy to be usable and assess how well it meets the user needs.

The MUSiC method provides the means to specify usability requirements and to measure how well those requirements are met. The MUSiC method can be applied, after a training course, without specific Human Factors or Human Computer Interaction experience and in a cost effective way.

2 The MUSiC Method

The MUSiC Method was originally developed as an ESPRIT project (MUSiC - *Measuring the Usability of Systems in Context* - ESPRIT Project 5429) partly funded by the European Commission. It consists of a set of methods and tools to evaluate the software quality of use (usability).

MUSiC allows to measure the extent to which a software product can be used effectively, efficiently and with satisfaction by specified users, for specified tasks in specified environments.

The MUSiC Measurement Method essentially provides two approaches to the evaluation:

1. the User-Based approach analyses the behaviour of users while completing some tasks in an appropriate environment where tasks, users and environment characteristics match those for which the product is being designed;
2. the Analytic or Theory-Based approach does not require the direct intervention of users because it analyses a model of the user-system interaction.

The MUSiC Method is composed by four modules which can be used independently or in an integrated way.

Three of the four modules are based on the User-Based approach. These modules are:

- User Satisfaction Measurement Method
- User Performance Measurement Method
- Cognitive Workload Measurement Method.

The fourth module, the Analytic Measurement Method, is based on the Theory-Based approach.

The four modules give measures of different aspects of usability and can be used at different stages of the software life cycle. Each of the modules is supported by software tools to collect and analyse the evaluation data (Fig. 1.). MUSiC states that it is not meaningful to talk about the usability of a product leaving the context in which it is used out of consideration [8]. A product can be considered usable only in relation to all the specific conditions under which it is used. For example a product which is usable by trained users may be unusable by untrained users. It is therefore necessary to study carefully the context of use of the product (the users, the tasks and the environment) and to relate the evaluation results to its context. MUSiC puts a strong emphasis to this aspect and all the four modules base the evaluation on a preliminary study of the context of use. The context study is supported by the questionnaire UCA (Usability Context Analysis) that allows to specify all the relevant aspects of the context from the usability point of view.

Method	*Software LifeCycle*	*Tools*	*Usability Aspects*
User Satisfaction Measurement Method	applicable from the stage “high fidelity prototype”	UCA SUMI SUMISCO	- Global usability - Efficiency - Control - Helpfulness - Learnability - Affect
User Performance Measurement Method	applicable from the stage “low fidelity prototype”	UCA DRUM	- Effectiveness - Efficiency - Relative user efficiency - Productive Period
Cognitive Workload Measurement Method	applicable from the stage “high fidelity prototype”	UCA SMEQ TLX Vitaport	- Task Performance (subjective and objective) - User Effort (subjective and objective)
Analytic Measurement Method	applicable from the stage “specification”	UCA SANe miniSANe	- Learnability - Cognitive Workload - Effort for error recovery - Efficiency - Adaptedeness of the system to user procedures

Fig. 1. The MUSiC Methods and Tools related to the stages in software lifecycle

2.1 The User Satisfaction Method

The User Satisfaction Method utilizes the SUMI (Software Usability Measurement Inventory) questionnaire to assess the user perceived software quality [7]. SUMI is a 50-item questionnaire users, after a work session with the product, fill in. At least 10 representative users are required to obtain accurate results with SUMI.

The collected data can be analysed by the SUMISCO (SUMI SCOre) software tool that automatically derives the usability measures. The method can be applied when at least a prototype version of the product is available. The SUMI measures are indicators of the following usability aspects:

- Global usability
- Efficiency
- Control
- Helpfulness
- Learnability
- Affect.

2.2 The Performance Measurement Method

The Performance Measurement Method is based on the observation of the users while completing typical tasks using the product [13]. During the work session are gathered some measures related to the duration of the task, time spent in consulting help facilities and manuals, time spent in unproductive actions, amount of achieved task goals. The data collection can be supported by video recording equipment. The software tool DRUM (Diagnostic Recorder for Usability Measurement) automates many aspects of the video analysis process and derives automatically the user performance measures. The method can be applied when at least a prototype version of the product is available. The performance measures are indicators of the following usability aspects:

- Effectiveness
- Efficiency
- Relative user efficiency
- Productive Period.

2.3 The Cognitive Workload Measurement Method

The Cognitive Workload Measurement Method allows to produce subjective and objective Cognitive Workload measures [1]. The subjective measures are derived by the TLX (Task Load Index) and the SMEQ (Subjective Mental Effort Questionnaire) questionnaires the users fill in after a work session. The TLX is composed of six subscales related to the cognitive workload demand imposed on the users by the task. The users assign a rate to each of the six subscales. The SMEQ contains just one scale related to the effort invested during task performance to which the users assign a rate. The objective measures are heart rate variability and the respiration frequency. The heart rate variability is measured by applying three electrodes on the users' chest while respiration is measured by a transducer held in place by a band attached around the chest and the abdomen of the user. The method can be applied when at least a prototype version of the product is available. The cognitive workload measures are indicators of the following usability aspects:

- Task Performance (from a subjective and objective point of view)
- User Effort (from a subjective and objective point of view).

2.4 The Analytic Measurement Method

The Analytic Measurement Method derives the usability measures by the analysis of a dynamic model of the user interface and user tasks [5]. The building of the models is supported by a software tool (miniSANe or SANe - Skill Acquisition Network - for larger and more complex systems) that allows the simulation of the user procedures and derives automatically the analytic measures. The model can be built when the system specification are available. The analytic measures are indicators of the following usability aspects:

- Learnability
- Cognitive Workload
- Effort for error recovery
- Efficiency
- Adaptedeness of the system to the user procedures.

3 An Industrial Application of the MUSiC Method

Data Management experienced the use of the MUSiC method to evaluate the usability of different software products [4]. The modules were used both in an integrated way and separately in respect to the objective and the available resources for the evaluation. What follows is the description of an usability evaluation where all the MUSiC methods were applied.

3.1 The Evaluation Objectives

The evaluated product is a software product to manage all the problems related to the management of personnel in an organization. The product typical users are Senior Personnel Administrators and Secretaries. It works on (networked) advanced PC. At the moment of the evaluation, a migration from a first version of the product to a second version was implemented: the user interface of the product was converted from a semi-graphical interface to another one using icons, mouse and windows. Furthermore more functionalities were added, mainly for the analysis of personnel.

Some users, using mainly the DataEntry functions of the product reported difficulties with the second version, while they appeared to prefer the simpler character-oriented User Interface of product version 1.

In addition, users reported as difficult the concept of navigating among many objects before identifying on the screen the information required for some operations, mainly the evaluation functions.

The evaluation objectives were identified by the product manager and by the project manager.

The questions expected to be answered by the MUSiC evaluation of the product were the following:

A. Is the effort for changing the User Interface being allocated in the right direction?

B. What are the weak points of the various functionalities of the second version of the product, considering the problems described by users?

3.2 The Evaluation

The evaluation procedure was essentially of two different types depending on the two MUSiC measurement approaches: the Theory-Based measurement approach and the User-Based measurement approach.

The two kinds of evaluation took place in different physical and temporal sites.

3.2.1 The evaluation procedure for the Theory-Based measurement approach

In the case of the Theory-Based measurement approach the evaluation procedure was the following:

1. the objective of the evaluation were defined
2. the Context of Use of the product was defined
3. the device and the task model were built
4. a user-procedure simulation was performed, generating the user procedure automatically and printing the relevant measures
5. the measures were interpreted comparatively, considering the goals of the evaluation and the information gathered by the context study.

The two versions of the product, the semigraphical one and the windows based one, were compared.

The device model and the task model were built using the available documentation of the product and observing the User Interface of the evaluated product running on the PC.

3.2.2 The evaluation procedure for the User-Based measurement approach

In the case of the User-Based measurement approach the evaluation procedure was the following:

1. the objective of the evaluation were defined
2. the Context of Use and the Context of Measurement were defined
3. the User-Based session was performed
4. the obtained measures were interpreted comparatively, considering the goals of the evaluation and the information gathered by the context study.

Expert versus novice users were compared.

The evaluation session took place in an usability laboratory. This laboratory was provided with the normal equipment (video cameras, video recorder, etc.), with some additional instruments specific of the method (Macintosh), and with the material necessary to run the evaluated product (PC).

The users were chosen in order to reflect the characteristics of the planned users of the product. The participants were required to be experienced with the product but not to have specific experience of the tasks. The evaluation team asked them information about their previous experience with similar product, software packages, keyboard layout and the use of mouse.

The tasks were defined together with the development and the customer assistance team. They reflected the everyday tasks and totally took no more than one hour to be performed. They were of medium-high difficulty. The tasks were defined in order to avoid any kind of influence on ongoing situation.

3.2.3 The Context of Use of the evaluated product

According to the MUSiC Method, the Context of Use and the Context of Measurement of the evaluated product were defined and the following documents were produced:

- Product Questionnaire
- Context of Use
- Critical components which could affect usability
- Context of Measurement.

These elements provided a description of:

- the characteristics of the product with particular reference to those that were likely to affect usability
- the characteristics of the tasks for which the product was developed
- the organizational, technical and physical environment in which product is intended to be used
- the characteristics of the participants in the evaluation activities
- the evaluation tasks
- the organizational, technical and physical environment in which the evaluation took place.

The collected information were used during evaluation data analysis to better interpret the results.

3.2.4 Metrics applied

Considering the usability questions about the product, the metrics provided by the Theory-Based approach were used to evaluate usability between the first and the second version of the product.

The metrics provided by the User-Based approach (Performance, Subjective Cognitive Workload and User Satisfaction metrics) were applied only to the second version of the product.

3.2.5 Diagnostic measures applied

A number of issues were identified prior to the evaluation session. In order to ensure that the maximum amount of information was obtained for the session, a number of additional procedures were put in place as follows:

- Notes were made during the usability session.
- At the end of the evaluation session the users were interviewed about the difficulties they found by the operators.
- During the analysis of the tapes some markers were put to highlight the most common problems encountered by the users.

In order to identify if there were differences between the four tasks the following procedure was followed:

- the Subjective Mental Effort and the Task Load Index questionnaires data were analyzed; the differences between the scores for each task were identified as differences between task difficulties.
- the Performance Measurement Method was applied to each subtask.

3.2.6 Test sequence

The sequence of events in the User-Based usability session was the following:

1. The users had to seat at the desk and were given the first task instructions; they were informed about the use of the manuals, the necessity to call the operator immediately after having finished the task and about the existence of the task time limit.

2. After having finished the task or after the time limit has occurred (the time limit was decided to be a quarter of an hour considering that an expert user could perform the task maximum in five minutes.), the users were asked to fill in the Subjective Mental Effort Questionnaire (SMEQ) and the Task Load Index questionnaire (TLX).

3. The users performed the second, the third and the fourth task after having respectively received the task instructions and filled in the questionnaires after each task.

4. After the fourth task the users were asked to fill the SUMI questionnaire.

5. As the last event the users were asked to formulate their impression about the difficulties encountered during the task performance, about the causes of these difficulties and they were interviewed to obtain more data about their professional skills.

3.3 The Results of the Evaluation

The results of the evaluation were of two types: metric results and diagnostic results.

3.3.1 Analysis of the metric results

The analysis of the metric data produced the following answers to the usability questions about the product:

Answer to question A

Considering the existence of the two product user types, the first question can be answered as described in the following.

In spite of a difficulty to compare two systems which have been changed both in respect to user interface and functionality, it can be clearly stated that the enlarged functionality of the product version 2 benefits mainly the tasks of the managers, which use the functionality of the product to design courses and qualification measures. This benefit comes at the expense of a higher complexity for performing the data entry tasks. A trade-off exists, when comparing these two versions, between higher performance or lower decision complexity. The trade-off appears to be in favour for the product version 2 for the manager group of users, but seems to be unfavourable for secretaries.

Answer to question B

Questions relating to frequent navigation and resizing of windows can be answered, considering the existence of the two types of the product users. Part of the problem is caused by the necessity to resize and select windows. Assuming that a design can be found where this resizing operation is unnecessary, the advantage is higher for managers than for secretaries. The results show that the large number of commands, which are not required for data entry, but which are available at all times, makes the system difficult to use for secretaries.

Many problems related to the navigation among windows or to the use of the icons, have to be linked to the fact that the users find the organization of the interface not logical. It seems that there is no homogeneity in the product menus so that the users do not understand where and why they can perform some action or find some function. For example, sometimes the system allows them to drug icons, or to type fields, or to save changes simply by setting predefined values; other times it does not. In this way the users get lost and can not understand the general way the system operates. This “modularity” of the system is not supported either by an on line help or by an effective manual. So the system appears complex and uncaring and does not support the tasks the users have to do. A bit of reflection is needed if we think that the typical users should be both managers and secretaries. In fact, even if managers may be familiar enough with the use of complex systems, it’s sure that secretaries will find many difficulties in using such kind of product.

3.3.2 The Diagnostic Results

The data gathered by the evaluation session and during the evaluation session itself have offered the opportunity to produce some diagnostic data. In fact it has been possible to define problem areas and to obtain some additional information from the gathered data analysis not directly included in the method.

The diagnostic results consist of a set of twelve recommendations: seven for general problems and five for function specific problems. The specific recommendations are not reported here. About the general recommendations, just a short description of the solution is given below to give the readers an idea of the identified problems, without entering in the technical aspects of the suggestions.

General Problems:

- Lack of awareness / training
  Suggestion: the product is too complex to expect relatively novice users to use effectively with only limited training. It will be necessary a training course to be developed and sold as part of the product package.

- Poor manuals
  Suggestions: produce an index to the manuals formatted under headings representing user tasks and operations; produce a new user guide which is formatted under headings representing the operations and procedures that users wish to undertake, rather than formatted according to system functions or architecture.

- Difficulty locating functions
  Suggestions: make the menu names more expressive; provide a summary of the functions contained in the various menus in diagrammatic form in the user Guide or in on-line help.

- Inconsistent window scrolling
  Suggestion: for consistency all windows should have the same scrolling features.

- Quitting system in error
  Suggestion: add a secondary message asking if the user really does want to leave the product. The evidence is that users can misunderstand the first message.

- Using the click/double click of the mouse
  Suggestion: improve the system reaction to the mouse click/double click.

- Typing fields
  Suggestion: provide feedback message indicating how to fill the field in.

3.4 Cost/Benefit Analysis

In this section we expose an analysis of the costs against the benefits for the testing activity described above [3] [6]. Also, the costs and benefits of the subsequent modifications up to now are considered.

3.4.1 Costs

The costs of the evaluation were the following:

1. Time spent by the users: 30 p/d
2. Equipment for the usability evaluation: 1,800 ECU
3. Time spent for preparing the evaluation: 10 p/d
4. Time for the analysis of the metric data: 10 p/d
5. Time for training of evaluators: 11 p/d

3.4.2 Benefits

Two kinds of benefit have to be considered:

1. Increased sales or revenues
2. Decreased costs.

About the increased sales, this year the increment was about of 140%. The estimated contribution of the modifications suggested after the evaluation is about 20%. i.e. the incremented usability helped product sales for 20%.

At the same time, the project team reported a decreasing number of requests for assistance compared to the same period in the last year. The estimated decreased cost for assistance is 10 p/d.

Another impact of the modifications is a different and easier training course which resulted in a reduction in the cost for training of 10 p/d.

3.4.3 Cost/Benefits Assessment

After having separately considered costs and benefits of the MUSiC evaluation, we can now discuss the trade-off between them. The cost/benefit analysis is reported in the following table:

Costs

Effort	(61 p/d at 277 ECU)	16897 ECU
Equip.		1800 ECU
Total		18697 ECU

Benefits

Increased annual sales	33334 ECU
Decreased customer support	5540 ECU
Decreased training	5540 ECU
Total in one year	44414 ECU

Revenue 237.5%.

The cost of the effort for the corrections that were made as a result of the usability testing was not taken into account. It was because it always happens that, after some problems with users are identified, it is, of course, necessary to correct the faults. If these faults are identified without a structured test, but basing just on the implementors experience, it happens that the corrections are often not suitable to the problems resulting in a increased and less effective effort for correction.

4 Conclusions

The application of the four MUSiC methods allowed all the questions about the product to be answered, giving diagnosis and recommendations for its problems.

All the constraints related to the resources available for the evaluation (number of users, skill of the users, users' time, etc.) did not compromise the performance of the evaluation and were satisfactorily overcome.

The application of all four methods allowed to look into all the usability aspects about the product and each method gave significant results by itself.

After a short training course on the MUSiC methods, Data Management staff was able to use all the MUSiC tools without any problem and succeed in conducting the evaluation session without the help of experts in Human Factors.

The experience of a direct contact with users allowed people involved in the evaluation to understand better the existence of the gap between developers and users, resulting in an improved capability to communicate each other.

The results of the evaluation were so significant that some modification of the product were implemented immediately and others were planned. This caused an increment of the sales and a decreased effort for training and support, resulting in a favourable cost/benefit analysis.

5 References

1. Arnold A.G., Wiethoff M, Houwing EM (1991) *The value of sychophysiological Measures in Human-Computer Interaction.* In: Bullinger (1991).

2. Bevan N (1991) *Standards relevant to European Directives for display terminals.* In: Bullinger (1991).

3. Bias R.G., Mayhew D. J. (1994) *Cost-Justifying Usability.* Academic Press Harcourt Brace & Company Publishers.

4. Binucci L., Caracoglia G. (1993) *Usability Evaluation by MUSiC: An Industrial Application.* 1st European Conference on Software Testing Analysis & Review, Papers Volume.

5. Bosser T, Gunsthovel D (1991) *Predictive metrics for usability.* In: *Proceedings of the 4th International Conference on Human Computer Interaction* Stuttgart, September 1991. Elsevier.

6. Bullinger (1991) *Proceedings of the 4th International Conference on Human Computer Interaction* Stuttgart, September 1991. Elsevier.

7. Corbett M, , Porteous M, Kirakowski J(1992) *How to use software usability measurement inventory: the user's view of software quality.* Proceedings of the European Conference on Software Quality, 3-6 November 1992, Madrid.

8. Dillon A Macleod M, Maguire M, Maissel J, Thomas C, Rengger R, Sweeney M, (1993) *Context Guidelines Handbook* Version 3. National Physical Laboratory, Teddington, UK.

9. ISO (1993) ISO DIS 9241-11: *Guidelines for specifying and measuring usability.*

10. ISO 9126 IT - *Software product evaluation - Quality characteristics and guidelines for their use.*

11. Nielsen J. (1993) *Usability Engineering.* Academic Press.

12. Preece J., Rogers Y., Sharp H., Benyon D., Holland S., Carey T. (1994) *Human Computer Interaction.* Addison Wesley Publishing Company.

13. Rengger R, Macleod M, Bowden R, Blayney M, Bevan N (1993) *MUSiC Performance Measurements Handbook.* National Physical Laboratory, DITC, Teddington, UK.

Recent Industrial Experiences with Software Product Metrics

Ger Bakker, Fred Hirdes

TechForce, Hoofddorp, The Netherlands,
e-mail: Ger.Bakker@TechForce.nl

Abstract. We describe some project experiences using software product metrics in the second half of 1994. The goals of the projects were quality assurance and assessment, preventive maintenance test planning, migration, risk analysis, re-design, re-structuring, reverse engineering, and generation of new documentation. The clients were large organisations in railway, banking, insurance, government (taxing), social security, and telecommunication. We analysed > 10 million lines of code in COBOL, PL/I, Pascal, C, C++, and RPG. We report initial observations from these projects. In the future the results of these experiences will be analysed statistically. Some interesting results are already appearing at this moment: (i) problems in legacy systems have their cause in the design, not in the structure of the code. We see the consequences of years of maintenance of the software in the detoriated quality; (ii) re-design, and re-structuring of existing systems are less costly and safer solutions to these problems than re-building.

1. Introduction

At TechForce we specialise in software analysis, under the name COSMOS we deliver tools, services, consultancy in the area of software quality, and re-engineering. Most of our services are delivered in co-operation with partner companies, mostly software houses. COSMOS is developed by TechForce as a result of the ESPRIT project 2686 COSMOS. In this project software metrics were used in industrial trials, mainly at the sites of the partners Alcatel and British Telecom. From 1993 TechForce has re-engineered the COSMOS Metrics Workbench from prototype quality to industrial quality. New metrics were developed and implemented, especially for the higher level of abstracting of software systems. In 1994 COSMOS was extended with a re-engineering workbench. TechForce is continuing the research and development in the field of software metrics, and re-engineering. Present activities are metrics for software design, metrics for OO, and reverse engineering on higher levels of abstraction.

1.1 The goals of the projects

There are many reasons why the management of an organisation decides to investigate the existing software systems. Legacy systems are often considered as a threat for the organisation: old functionality, difficult to change, text oriented user interface, in one word old fashioned. However, a recent investigation of the Dutch taxing automation centre revealed some interesting results concerning the different view that users and IT personnel have on their legacy systems. Both groups filled out a questionnaire: most of the users of the old systems were satisfied with the functionality, and had confidence it them. On the other hand the IT personnel considered the systems as unacceptable to modern standards.

In more and more organisations, especially in banking, where you can expect a strong financial view, the software systems are considered a valuable asset, that are worth to be preserved. Often the investments in software development and maintenance over the last thirty years are impressive.
An argument for preserving the existing systems is the risk that the development of a new system poses. E.g. how to ensure that at least the old functionality is implemented in the new one? The manual system is not present anymore, and for that reason it cannot be analysed to model the business processes. The development of a completely new system is still very expensive, and time consuming, and CASE has not delivered as sometimes was hoped.
Three are many other reason to analyse the existing systems. To mention some: migration to another hardware or software platform, change to relational data base technology, quality assessment, preventive maintenance, re-documentation, inventarisation, and many others.

The next table gives an impression of the diversity of the goals and environments of our recent work.

Organisation	*Goal of the project*	*Language*	*Environment*	*Application*	*Size Mloc*
Banking	quality assessment	COBOL	IBM	ATM	0.1
Railway	code review	Pascal	DEC/VAX	control	0.2
Social security	acceptance	COBOL	DEC/VAX	sick fund	0.4
Banking	inventarisation	COBOL, PL/I	IBM	several	0.8
Insurance	risk analysis	RPG	IBM	several	1.3
Government	reverse engineering	COBOL	IBM	taxing	0.6
Insurance	conversion	COBOL	Bull	several	1.5
Telecom	acceptance testing	C++	PC	billing	0.3
Telecom	maintenance	C	Tandem	managem.	0.6
Banking	migration	COBOL	Bull	all appl.	2.0
Whole sale	data model recovery	COBOL	DEC/VAX	all	1.9
Insurance	maintenance	COBOL	IBM	several	1.1
Cosmetics	maintenance	RPG	IBM	several	0.5
					11.3

Table 1. The metrics projects during the second half of 1994.

1.2 The backgrounds: graph theory, metrics, GQM and ISO 9126

The use of software metrics is not common practice in industry yet. Our clients mostly decided to have the analysis executed in service by an outside specialist company. We expect that this will change to a situation where the organisations will have acquired metrics expertise, have their own tools, and perform the analysis

themselves. This is already happening in the world of technical systems, business systems will follow.
The application of software metrics is not a trivial activity: there are many pitfalls. One needs to investigate goals, to get commitment, to make a plan, and to execute it. In the end results must be delivered that are meaningful for the different levels of the organisation: managers, users, and IT professionals. One should not forget that metrics often mean the delivery of bad news. So also the communication of goals and management commitment is important. Metrics may not be used to judge people.

To make these projects deliver valuable, reliable and objective information to the client one needs to apply scientifically and technically sound principles. The COSMOS approach is based on the Dijkstra prime theory, and the work of [11]. Stainer [8] added the descriptor theory to the flow graph. By this generalisation we were able to also include design models and parallel processing in our model. After the COSMOS project we replaced the descriptor tree with an object oriented model [4], [15], [13]. The current model can contain virtually all semantic information of a software source text, and is suitable for the classic imperative languages, parallel processing languages, formal specifications, and design languages. The software metrics we use are part validated in the COSMOS project [11], some are from literature [7], [3], [14].
At TechForce we use the results of the ESPRIT project [1] as a guideline for the analysis process. The AMI approach and the ISO/IEC 9126 standard form the base to design our measurement plan. More about the analysis proces in the next paragraphs.

The AMI approach uses twelve steps in what appears to be a Deming circle: Assess, Analyse, Metricate, Improve.

Application of Metrics in Industry	
Assess	1. Environment 2. Primary goals 3. Validate goals
Analyse	4. Break down into sub-goals 5. Check consistency of the goal tree 6. Produce table of questions
Metricate	7. Write the measurement plan 8 Collect the primitive data 9. Verify the primitive data
Improve	10. Distribute, analyse, review the data 11. Validate metrics 12. Relate the data to goals and implement actions

Table 2. The steps of AMI

For each projects we make a measurement plan that serves two main goals:

- for the client it shows the relationship between business goals, goals for the IT, and software metrics.
- it is the input for the TechForce people that execute the actual measurements

To illustrate the application of AMI table 2 shows an excerpt of a measurement plan with code review as its main issue. This project involved the process improvement for the development of railway control systems. We implemented software metrics for code and design review.

	Question	Remarks	Product/metric	Definition
1	Naming	Check on standards as mentioned in [1], § 7.2.6. De standaard geeft de mogelijkheid tot het opnemen van een project extensie in de naamgeving. Hiervoor zal geen metriek worden geleverd. The § 7.2.6, p. 45, 1e "*" standard concerning the lettercode prefix and § 7.2.6, p. 44, 3e "*" : occurence of a prefix will be checked	MNAME	# violations
2	Layout	Not implemented in the pilot		
3	Comment	Metrics : the amount of comment, the deviation. See also [1], § 7.2.3. Comment at the e.o.l. is not considered to be comment. Comment defined as : each line without valid Pascal elements on the same line, where comment consists of at least 3 words of at least 2 characters. The comment metrics is given for each subroutine. Comment before a subroutine is considered to belong to the subroutine. A correlation diagram between lines and comment is given. Via interactive COSMOS it is possible to view the quality of the metrics.	MCMT CFCMT Diagraml	# comment lines Correl. diagram
4	Conform design	Not directly performed by COSMOS. See remarks after table..		

Table 3. Part of a measurement plan.

2. The instrumentation

To analyse the source code we use our COSMOS Metrics Workbench, and COSMOS Re-engineering Workbench. The first is used to analyse the population of software components using statistical analysis of software metrics. The second is used to analyse individual modules to retrieve more insight it the how and why of metrics results. These explanation uses graph theory, and several graph based presentations.

The tools run on standard UNIX workstations, we use SUN Sparc and HP9000 series 700. This requires all software to be transported in ASCI code to the UNIX environment. This is no problem in most cases using diskettes or e-mail for the smaller projects, LAN solutions for the large systems.

COSMOS Architecture

COSMOS is built in a modular way to ensure flexibility, and to make it an open system that can be easily connected to other systems. Flexibility is especially important to make adaptations possible to the system under test. For instance for operating system features, language dialects, job control, DBMS features, and most important to the goal that has to be achieved with the analysis. COSMOS consists of three main modules.

Input	*COSMOS module*	*Output*
ASCII source code EDT model metrics numbers graphic representations	Language interface Core analysis engine & OTCL Management support interface	EDT model metrics numbers graphic representations metrics graphics statistics reports prints

Table 4. The COSMOS system architecture, and its (intermediate) products.

Languages

The basic prerequisite for COSMOS is some formal representation of the design or the software code, that can be translated in a formal grammar. Generally spoken it is possible to analyse with COSMOS any product that is stated in a formal language, or textual, or graphical. The state-of-the-art is that COSMOS can analyse several specification languages [8], [9], [5], [2] as well as implementation languages, job control, and database/data definition languages. Some examples of supported languages are:
LOTOS, SDL, COBOL, C, C++, PL/1, RPG, CICS, DL/1, Pascal, SQL, JCL
For several customers language interfaces for their proprietary languages have been developed. In general the development of a new language interface is a well-defined and straightforward process, and is facilitated by a generic set of internally developed tools. For the implementation of design metrics w e drew on the work of [6], [10], [12] and [9].

OTCL.

COSMOS comes with a standard set of metrics that is general for most of the supported languages, and some metrics that are language or problem specific. These metrics are defined using the COSMOS OTCL, which is also used to define project specific metrics.

2.1 The analysis process

The analysis proces consists of several steps.

- Measurement plan.
- Intake of the software.
- Measurement.
- Analysis of the results.
- Reporting.

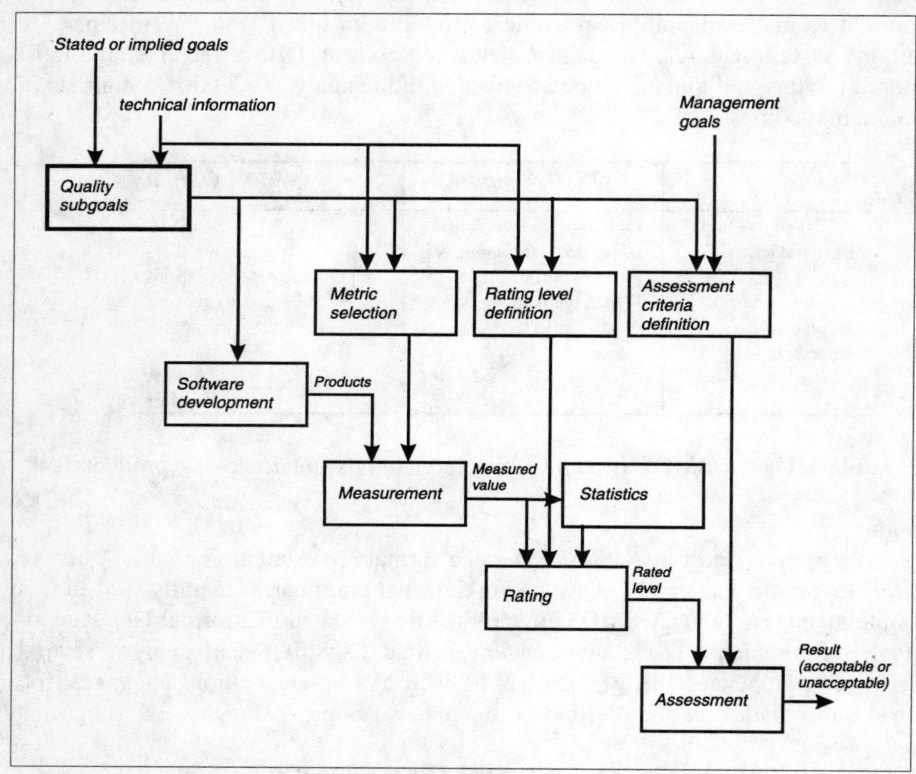

Fig. 1 Evaluation process model

2.1.1 Measurement plan

In the contract with the client the main goals of the project are stated. In the measurement plan these goals are further refined, and the goal, question, metrics tree is derived.

From the measurement plan the TechForce specialists create the specific scripts to generate the output, that is not in the standard set.

2.1.2 Intake of the software

In this phased the source code is collected and transported to the UNIX workstation. Sometimes we use e-mail, other times diskette, magnetic tape. For large systems we connect directly to the LAN of the client. In existing systems we encounter two aspects that complicate the intake of software. First it takes always some effort to make the system complete: source files are missing, copy members not complete or versions not clear. Second there is the problem of non-standard implementations of the languages, or the use of hidden features of certain compilers. In general we try to replace the statements that give syntactical problems with the correct ones, sometimes we have to adapt the parser.

2.1.3 Measurement

After the software is completed and is suitable for parsing, the measurements are performed. Depending on the results that have to be delivered this can be a one-step process or it can take several iterations. E.g. iterations are needed when variables must be traced.

2.1.4 Analysis of the results

We analyse the metrics results first, top-down: system level metrics, module level metrics, subroutine level metrics. In COSMOS we do not often work with pre-defined metrics levels for accepting or rejecting a software component, but instead we prefer to compare the population with itself. The components that are exceptions in the population are reported. This approach is for many people more acceptable, and easier to communicate , compared to a rating by "experts".
Normally we use boxplots for the statistical rating of the metrics, only when fixed rating levels are available or given by the client we also use these. One should be very careful using descriptive statistic on population with unknown distributions: we use the median instead of the average, and refrain from using standard deviation. The population view is derived from the metrics, if appropriate we look at individual components (sub-system, module, subroutine) using the re-engineering workbench. We generate call graphs (often known as structure charts), flow graphs, object hierarchies, data models and special reports. [16], [17].

2.1.5 Reporting

In the reporting phase the material is ordered, we create spreadsheets, sometimes do some additional calculation. When necessary the reports, metrics and graphs are collected in a machine readable format to be used by the client in a next step in a project.
We end the project with a presentation of the results for the client, sometimes accompanied with an on-line demonstration on the clients software

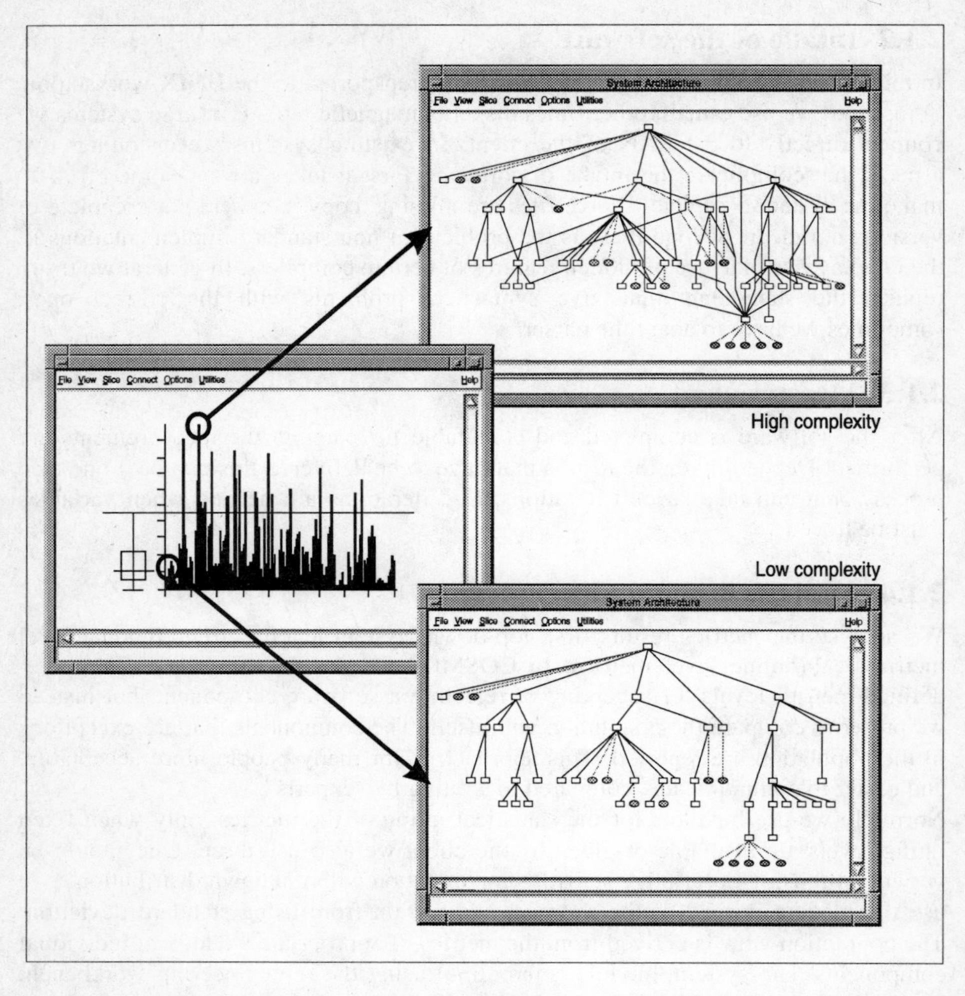

Fig. 2. How an architectural metric shows the difference between call graphs with a high and a low complexity.

.

3. Results

As stated before, we have no intentions yet to draw firm conclusions about the state of legacy systems in terms of statistical analysis.
Our results show that a lot can be learned from software metrics, they can be valuable for general management, IT management and the software engineers.

However for these groups, with their special interests, one should deliver results on the right level of abstraction:

- the general management is interested in an overall view on the quality of the system, in the cost of maintenance and the cost of change. Here we go not into more detail than a module (program), and not more than a few metrics.
- the IT management can expect to get in control of the development and maintenance process. Without really tightly inspecting the engineers one can have a good and objective view an quality, implement actions, and see the results. On this level it is possible to report parts of modules (subroutines), using several metrics.
- the software engineers can use software metrics to improve their own performance: better understanding leads to better code, but also higher productivity by easier change, better and shorter testing and reuse. By doing it themselves the costly action of corrective maintenance is not necessary anymore: first time right. Software engineers are normally only concerned with their "own" program, so much more detail can be delivered.

From the projects mentioned we have in general very positive reactions. We see that the results of our analysis is often used to create a platform of communication between managers and engineers. In many projects we had managers report to us that for the first time they were able to discuss software quality issues in a very concrete manner. An important gain in most projects was, that the discussion took place on objective criteria, delivered by a tool. For the engineers this took away the danger of be judged as a person.

For the reported projects we give some of the results we delivered.

Some impressions on architectural level are reported in Tab.5.

The impression on structural level is that here control over quality is better. It seems that software engineers have a better understanding of on the details (subroutines) than of the design. Mainly we analysed the products of large teams, where software engineers are not required to understand the whole system (see Tab.6).

4. Conclusions

From the point of view of metrics specialists we always are very curious about the overall state of the quality of legacy systems.
As said before we feel we have not yet enough results to make statistically sound conclusions. Also this would take must more time. Because there are some strong indications that we wanted to share with the IT community, we decided to report our results in an informal way first.

Organisation	*Goal of the project*	*Metrics results*	*Conclusion*	*Lesson learned client*	*Lesson learned by TF*
Banking	quality assessment	Low design complexity, one exception	good maintainability and testability	Improve one module	
Railway	code review	high design complexity metrics	Need better understanding of design	Take care of relation with technical design	Need to see interprocess communicatio n
Social security	acceptance	low design complexity, one exception	OK, rewrite one module	all releases must be measured	need job control analysis
Banking	inventarisation	technical function points			
Insurance	risk analysis	size metrics high, architecture large, not complex	OK, large programs, high, functionality	split modules	
Government	reverse engineering	very large, high design complexity	very difficult to maintain	remove old functionality, split modules	bad news is threat for messenger
Insurance	conversion	high design complexity metrics	complete inventory of changes	JCL needs management	time pressure is bad for quality
Telecom	acceptance testing	low design complexity, some exceptions need rework	very good system	reduce maintenance costs, replace old system	very high performance, but good maintainability
Telecom	maintenance	high design complexity	technical and functional changes very difficult	needs re-engineering	K&R C in the worst form
Banking	migration	high design complexity	migration cost statement	complex design, complicated by JCL	
Whole sale	data model recovery	low design complexity, small modules	ERD model	reuse is not always wise	hidden file/data access is difficult to trace
Insurance	maintenance	size and complexity metrics high	maintainability improvement proposal		choose the right metrics
Cosmetics	maintenance	high design complexity metrics	maintainability must be improved	re-documen tation	

Table 5.

Organisation	*Goal of the project*	*Metrics results*	*Conclusion*	*Lesson learned client*	*Lesson learned by TF*
Banking	quality assessment	low complexity and size	some subroutines need improvement of testability	in control of quality	testability problems yield high testing costs
Railway	code review	high complexity and testability metrics	need lower complexity, and testability	perform 100% code review	automated code review is very powerful
Social security	acceptance	simple subroutines			
Banking	inventarisa-tion	technical function points		no manual FP available on subroutine level	comparison with calculated FP is difficult
Insurance	risk analysis	simple subroutines	OK on structural level		a few bad routines created the havoc
Government	reverse engineering	high complexity,	very difficult to maintain	rewrite some subroutines	flow graphs are not common understanding
Insurance	conversion	high complexity,	complete inventory of changes		Cobol dialects can have strange behaviour
Telecom	acceptance testing	low complexity, some exceptions need rework	very good structure		need for better OO metrics
Telecom	maintenance	very complex, low testability	technical and functional changes very difficult	needs re-structuring	K&R C in the worst form
Banking	migration	low complexity subroutines	migration cost statement	concentrate on portability	refrain from using proprietary extensions
Whole sale	data model recovery	low complexity, small size	impact of change to RDBMS	refrain from using proprietary COBOL extensions	
Insurance	maintenance	size and complexity metrics high	maintainability improvement proposal		choose the right metrics
Cosmetics	maintenance	high complexity metrics	understandability is at risk	re-documenta-tion	flow graphs are powerful for RPG subroutines

Table 6.

Our main conclusions are these:

1. We see the consequences of years of maintenance of the software in the detoriated quality. However we see no strong evidence that re-structuring is the solution. We noted the important problems in the architecture, not in the structure. This misunderstanding is logical when ones view is limited to a few lines of the source code. Subroutine level metrics and flow graphs show that the programmers did a reasonable good job. Improvement of the structure of a subroutine is only of limited impact on the whole system. Improvement of the architecture makes functional adaptations possible, a pre-requisite for survival of the system.
2. Re-design, and re-structuring of existing systems seems to be a less costly and safer solutions to the quality problems than re-building, provided that the functionality is sufficient. We saw that on all levels of the systems only a smal part of the components were responsible for the quality problems. When considering the cost of replacing a system of several million lines of code with an entire new one, this is a very expensive solution to solve quality problems that have their source in only 5 to 10 % of the code. Measurement can place these decisions in the right perspective.

We saw the importance of sound design, and with that the importance of taking care of the design during maintenance. Often changes are implemented on the code level, but their implications on the design of a program are not realised. The importance of design for maintainability was also reported by [6]. Detoriated quality of subroutines is in our experience easier to correct than the detoriated quality of the architecture of a program, or even worse, a system. However we used the clustering algorithm of the COSMOS re-engineering workbench to solve some “spaghetti design problems”quite succesfully.
For the second point we saw that in all software systems their is a 80/20 rule. Of all the programs in a system only (roughly) 20% is considered a quality threat. So 80% is of good or acceptable quality, and an asset for the organisation. The same is valid for the subroutine level: of all the programs considered for preventive maintenance, only 20% of the subroutines have structure problems. This means that to “rejuvenate” a system, it is often possible to do that without touching the functionality, and that it will have implications for a low percentage (the already mentioned 5 to 10 %) of the source code. The power of the analysis is that we can pinpoint the bad parts, and how to attack the source of the problem. It means that the costs of rejuvenation could be paid from the normal (adaptive) maintenance budget, and that it will pay back from the lower corrective maintenance costs. The problem is that maintenance budgets always have been allocated to functional changes. Software does not wear down, so organisations are not used to allocate budgets for “preventive” maintenance of software. We think we have a point to make that change in the future.

5. References

1. AMI Application of Metrics in Industry, A quantitative approach to software management, AMI.
2. Belina Ferenc, Dieter Hogrefe and Amardeo Sarma, SDL with applications from protocol specification, Prentice Hall International, New Jersey, USA, 1991.
3. Boehm, B.W., A spiral Model of software development and enhancement, IEEE computer 21 - 5, pp 61-72, 1988.
4. Booch G., Software Engineering with ADA, The Benjamin/Cummings Publishing Company Inc., Menlo Park USA, 1985.
5. Braek R., O. Haugen, Engineering real time systems, An object-oriented methodology using SDL, Prentice Hall International, New Jersey, USA, 1993.
6. Card, David N., Robert L. Glass, Measuring software design quality, Prentice Hall International, New Jersey, USA, 1990.
7. Conte, S.D., H.E. Dunsmore and V.Y. Shen, Software engineering metrics and models, Benjamin/Cummings publishing company, Menlo Park, USA, 1986.
8. Stainer Sieglinde, B. Kalipcoglu , State of the art information flow metrics, Alcatel, Austria, 11 december 1991.
9. Giang Don Hoang, Development and implementation of a frontend for the design method SDL on block layer, Alcatel, Austria, 10 october 1992.
10. DeMarco, Tom, Controlling software projects, management measurement & estimation, Prentice Hall International, New Jersey, USA, 1982.
11. Fenton, Norman E. Software Metrics, A rigorous approach, Chapman & Hall, London, England, 1991.
12. Grady Robert B., Software Metrics, Establishing a company-wide program, Prentice Hall International, New Jersey, USA, 1987.
13. Meyer Bertrand, Object-oriented Software Construction, Prentice Hall International, New Jersey, USA, 1988.
14. Möller, K. H., D. J. Paulish, Software metrics, A practioner's guide to improve product development, Chapman & Hall, London, England, 1993.
15. Rumbaugh, James, Michael Blaha e.a., Object-Oriented modelling and design, Prentice Hall, International, New Jersey, USA, 1991.
16. Youl, David P., Making software development visible, effective project control, John Wiley & sons, Chichester, England, 1990.
17. Yourdon, Edward, Larry L. Constantine, Structured Design, Fundamentals of a discipline of computer program and system design, Prentice Hall, International, New Jersey, USA, 1979.

Quality Measurement of Software Products: An Experience About a Large Automation System

Andrea Spinelli[1], Daniela Pina[1], Paolo Salvaneschi[1]
Ernani Crivelli[2], Roberto Meda[2]

[1] ISMES SpA, Viale Giulio Cesare 29, I-24124 Bergamo BG
tel. +39-35-307322, fax +39-35-211191, e-mail dpina@ismes.it

[2] ENEL/DSR-CRA, Viale Alessandro Volta 1, I-20093 Cologno Monzese MI
tel. +39-2-7224-5551, fax +39-2-7224-5548, e-mail filippi@cra.enel.it

Abstract. We present our experience in measuring quality of a large automation system. Our approach was to start from the state of the art in quality models, to formalise the expression of quality requirements, to build a quality matrix, which relates quality requirements to each single functionality, to apply several pruning techniques to cut down the measurements to be taken. Our approach allowed us to manage the complexities involved in quality measuring of large systems: it is difficult to express quality requirements, measurement costs are high, users need something more specific than a single quality profile over a complete application. We discuss a specific real case, with some practical implications. We conclude quoting some possible extensions.

1. Introduction

Quality evaluation of real real-life software systems requires a significant effort; in fact the literature describes many measurement procedures, based on quality models, and related measurement instruments, but they can't be used in practice without a significant theoretical and practical work. In essence, starting with a data base of measurement instruments, several specialisation criteria are required to adapt the available measurement environments, according to the evolution in time of a software system (documents), to its components (functionalities) and to the depth of the measurement.

Moreover, since evaluation is a costly process, for significantly-sized systems it is necessary to limit the breadth of the analysis to the most critical parts, according to well-defined criteria.

Finally, automated tools are needed to manage the complexity related to the execution of a procedure, specialisation according to given criteria, storing of measurements, and integration into synthetic values.

This article describes in some detail the experience resulting from practical application of the technology to some actual case studies: starting from a resume of the state of the art, we identify some problems in the existing measurement guide lines. We introduce some improvements and optimisations to current practices and we discuss a real-life case study.

Some topics for further research and an overall discussion conclude the article.

2. State of the art

The literature describes several **quality models** (see[1], [2]), which have had limited application, and which have had a standard expression in the ISO9126 document [3], [4]. They share a common structure, in which the user is presented with a set of values for some external characteristics - easy to understand from the user point of view, and those characteristics are related with internal characteristics - easier to measure from a technical point of view, using a set of measurement instruments.

Figure 1 depicts graphically the quality model and related measurement instruments proposed by [1].

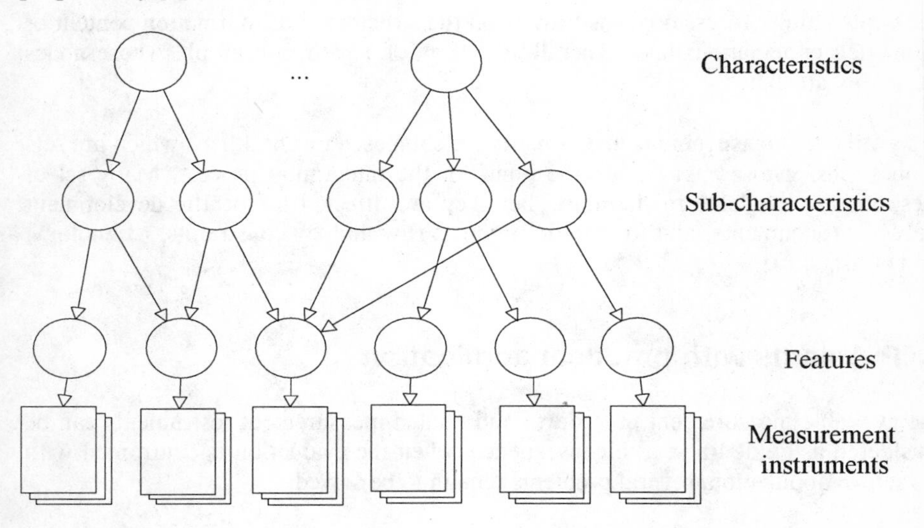

Fig. 1. Bowen's Quality Model

The **evaluation procedures** available in literature ([1], [5]) represent the quality model as a set of equation and parameters describing the relationship among measurable properties of a software item and a set of quality characteristics.

Generally an evaluation procedure defines three operations over a quality model:

1. *derivation*: given a required quality profile, generate a set of measures to be taken, trying to minimise measurement cost;

2. *integration*: given a set of measurements, generate a quality profile.

3. *explanation*: after an integration process, explain why the measurements produced that quality profile

The derivation process is based on the fact that it is useless to measure properties which are not interesting for the user. As such, it considers only the characteristics for which non-low value is required, and includes in the measurement plan only the measures which contribute to them.

The integration process is usually based on some form of weighted average or minimum operator among measured values. An important aspect of the integration process is the choice of interpretation functions, which map objective values (e.g. *the average cyclomatic number is 7.3*) onto merit values (e.g. *0.8 on a scale from 0.0 (worst) to 1.0 (best)*).

The explanation process depends heavily on the structure and information content of the model; explanations take generally the form of a path in a graph, whose nodes are quality attributes.

The available **measurement instruments** are composed by checklists, which are sets of questions, whose answers are the input for the integration process. Many set of questions are available in literature, but they are directed to specific development processes (documents) and to specific products (hw and sw constraints, technology, kind of users,...).

3. Problems with practical application

The available measurement procedures and related measurement instruments can be considered as guide lines; as a consequence, when the evaluation is confronted with a practical application, several problems remain to be solved.

First, the user must express the quality requirements in a formal way; of course, it is difficult to understand what are the real requirements and some elicitation methodology is required to provide an interpretation of the situation. The cost of this expression process must be viable in the specific domain.

Moreover, when considering complex systems, the unit of application of the quality evaluation process is not the whole system, but the single functionality (from a user point of view). For instance, reporting activities may be required to be maintainable, because they change often, but not as reliable as functionalities which have an impact on human lives.

Second, not all criteria are good for all environments: numeric computing software has, on the average, a higher cyclomatic number than, for instance, accounting software of comparable quality. As a consequence, the whole interpretation apparatus must be specialised for specific application domains. Moreover, each domain may have indicators which are specific for that domain; for instance, not all products need a user interface. In the specific case of the ISO characteristic *Functionality*, it is necessary to take into account the *implied user needs*, which means to define which are the normal expectations of a user in a given historical context; as a result, it is necessary to define which are the functionalities deemed necessary for a given application domain.

Finally, the cost of the evaluation must be reasonable. Depending on the size and the complexity of the product it could be so high to make unlikely the evaluation.

4. Improvements to the evaluation procedure

Our experience in measuring software quality results from practical applications in the fields of scientific computing tools and of large (>500K lines of code) industrial automation systems. In this section we explain the techniques we had to introduce in order to cope with the difficulties described before.

4.1. Expression of required quality profile

Current development methodologies provide tools to define functional requirements, but there is a lack of tools for the expression of quality requirements. Our approach is quite simple, but tries to fill the vacancy in a straightforward manner.

A quality profile is a set of values (we choose a [0,1] scale) for the ISO characteristics. First of all, we identify a general required quality profile by means of a questionnaire: looking at some attributes of the application, we derive a rough estimate of required values, in qualitative terms (High-Medium-Low).

For each characteristic, the questionnaire identifies some attributes, which are mediated and interpreted to give a synthetic value. Figure 2 quotes a fragment of the questionnaire dealing with usability.

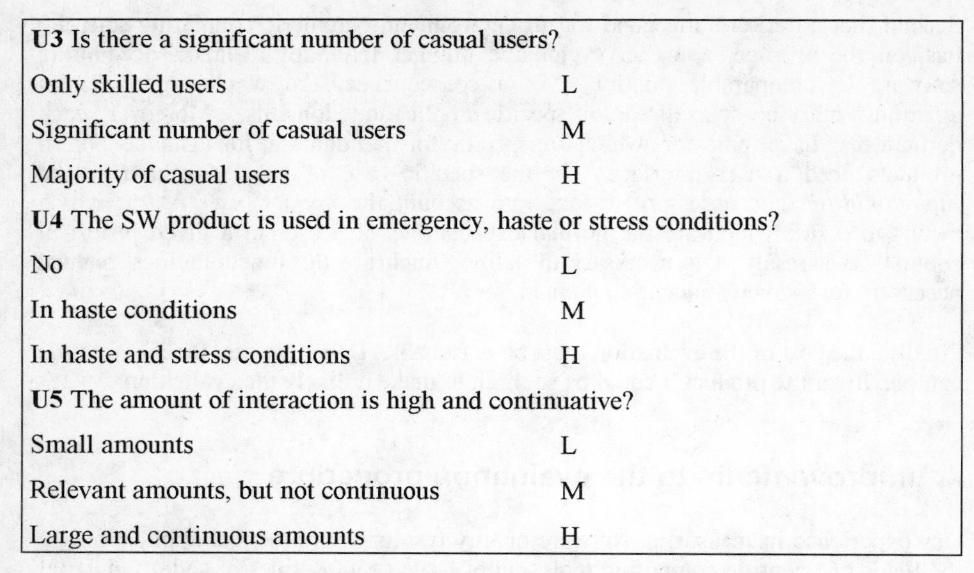

U3 Is there a significant number of casual users?

Only skilled users	L
Significant number of casual users	M
Majority of casual users	H

U4 The SW product is used in emergency, haste or stress conditions?

No	L
In haste conditions	M
In haste and stress conditions	H

U5 The amount of interaction is high and continuative?

Small amounts	L
Relevant amounts, but not continuous	M
Large and continuous amounts	H

Fig. 2. A fragment of the questionnaire dealing with usability

The general quality profile gives a high-level information on the product and might even be appropriate for small-scale systems. However, discussing with users and producers, we discovered the quality requirements are seldom uniform over a whole product. For instance, high efficiency may be required in the graphic display functionality, but not in the batch print functionality; reliability is necessary on mission-critical functions, but not on complementary components of the application; maintainability is more necessary in functionalities which are more likely to evolve over time.

In order to satisfy user need for requirements expression, we developed a two-stage methodology for required quality expression. In the first stage, a simple functional hierarchical model of the application is developed, using standard techniques. The whole system appears explicitly at the root of the tree. In the second stage, we lay out a matrix, whose rows are functionalities and whose columns are ISO characteristics, and, using again the questionnaire and some application domain knowledge, we elicit from the user any specific requirement for specific functionalities. In practice, it is not reasonable to ask for a granularity better than a High-Medium-Low classification.

Using a form of reasoning by default, when the user does not explicitly specify any requirement, we inherit the requirement from the next-higher requirement, and, if necessary, from the general quality profile.

The following table (Fig. 3.) is an example of quality matrix for a generic process-control application; in the first row, we have the general quality profile from the questionnaire; in the rows below, we have the quality requirements for each individual function. The highlighted valued are those explicitly required by the user.

ID					EF	FU	MA	PO	RE	US
m	System				L	M	M	L	H	H
m1	Controlling system				L	M	M	L	H	L
m1.1		Startup			L	L	M	L	H	L
m1.2		Commuting			**H**	H	M	L	H	L
m1.3		State transition			L	M	M	L	H	L
m1.3.1			Automatic		**H**	H	M	L	H	L
m1.3.2			Manual		L	M	M	L	M	H
m2	Controlled system				L	M	M	L	H	H
m2.1		Acquisition			L	M	H	L	H	H
m2.1.1			Manual		L	M	H	L	M	H
m2.1.2			Automatic		M	M	H	L	H	L
m2.2		Model			L	**H**	M	L	H	H
m2.2.1			Editing		L	H	M	L	M	H
m2.2.2			Use		**H**	H	M	L	**H**	H
m2.3		Diagnostics			H	**H**	H	L	H	L
m2.3.1			Sensors		H	H	H	L	H	L
m2.3.2			Actuators		H	H	H	L	H	L
m2.3.3			Communicati		H	H	H	L	H	L
m2.4		Commands			L	H	M	L	H	L
m2.5		Comp/Human IF			L	M	M	L	H	H
m2.5.1			Operators		M	M	M	L	H	M
m2.5.2			Model display		L	H	M	L	H	H
m2.5.2.1				Screen	**H**	**H**	M	L	H	**H**
m2.5.2.2				Ascii print	L	M	M	L	L	M
m2.5.2.3				Plotting	L	M	M	L	L	M
m2.5.3			Commands		H	H	M	L	M	H
m2.5.4			Alarms		H	H	M	L	H	**H**
m3	Services				L	M	M	L	H	H
m3.1		Journaling			L	L	M	L	M	L
m3.2		Backup & rec			L	H	M	L	H	M
m3.3		Data imp t/exp			L	M	M	L	L	M

Fig 3. An example of quality matrix for a generic process-control application

The final output of the process is a *quality matrix,* which associates with each functional component a quality requirement. The quality matrix is used as a guide for the measuring process.

The quality matrix has proven successful in order to satisfy some clear user needs; its general applicability becomes clearer when contrasted with cost estimates.

As to the costs, the general quality profile expression is cheap, requiring on the order of one work day to complete. Expression of the functional model and expression of quality requirements may require some more work (about one work week), but, above all, needs some co-ordination with the functional requirements specification.

Quality matrix formulation is the first case that explicitates the need for a classification of applications in homogeneous classes. Each application in a class shares with the others a set of common functionalities and maybe a common set of techniques and technologies. For instance, the class *Programming editors* shares functionalities such as *Open File*, *Save File*, *Search and Replace* and so on. An interested party may wish to write down a common requirements specification and a common quality matrix for all programming editors, dividing the effort over all members of the class. The need for classification will become more evident later.

4.2. Organisation, specialisation and management of the measurement apparatus

In order to build an effective measurement apparatus, we proposed a schema on the formulation of all measurement instruments, as in the following example (Fig. 4.):

Id: 201
Source: Bowen
Feature: ***Documentation clarity and information content***
Question:
Do software requirements and architecture have separations based on functions?
How:
Find a suitable list of functions (if not present, answer NO); cross-check the list with the separators.
Notes:
Check that the same function names are used in both cases
Tools: *Document visual inspection*
Assessment level: 2
Interpretation function: YN: yes $\rightarrow$ 1, no $\rightarrow$ 0
Applicable documents:
Software requirements, general design, detailed design

Fig. 4. An example of measurement instrument

Apart from unique ID and source, each instrument is characterised by the feature it measures, the main question, and several indications about how to measure and how to use the result. Specifically, we outline what documents must be searched, how the measure has to be performed, optional clarifications, what tools must be used. The instrument is related a standard *interpretation function*. Strictly speaking, the interpretation is more related to the quality model, but we included it here because it makes clearer the purpose of the question.

Taking for granted a [0,1] scale for the merit, where 0 is worst and 1 is best, possible examples of interpretation functions are the *positive Boolean* (yes=1, no=0), *negative Boolean* (yes=0, no=1), *linear* (if the answer is a number x, take x as the output), *quadratic* (if the answer is x, take x^2 as interpretation), *hyperbolic* (if the answer is x, take 1/x as the interpretation).

In order to obtain a measurement apparatus usable in different kinds of applications, we specialised it using three different levels of precision; we devised a schema with three *levels of assessment* (fig. 4.), defined as follows:

1. level 1: instruments evaluating the measurement applicability.

2. level 2: instruments with the purpose to verify general properties of software from a perspective of good software engineering practice.

3. level 3: instruments related to specific technologies and to specific attributes related to functionalities.

The first level points out: possible macroscopic deficiencies in the project management or leadership, real viability of the evaluation and some indications about the related cost.

The second level of assessment gives a much more detailed account of the status of the system, guaranteeing a better precision of measurements.

The third level requires the construction of the quality matrix and of a set of specialised indicators, linked with the specific functionalities and technologies entailed by the application characteristics.

Our measurement apparatus includes about 550 such instruments. Of course, it is not viable to handle them manually. We developed a tool, IQUAL, which main functionalities include:

1. management of multiple quality models
2. management of multiple measurement apparati
3. derivation of measurement plans
4. management of measurement results
5. integration of measurements into quality profiles
6. explanation of profiles in term of the quality model

The current version is described in [6]. We are currently developing version 4.0, which will fully exploit the Windows environment. An hardcopy of the page related to the above mentioned measurement instrument is shown below (Fig. 5).

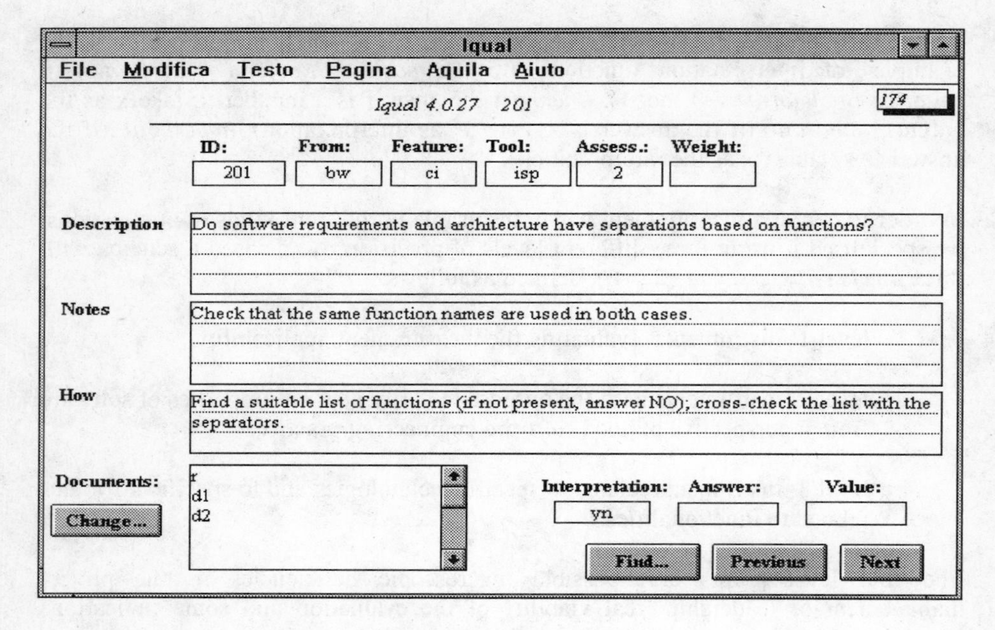

Fig. 5. IQual: an hardcopy of the screen

4.3. Reducing the measurement effort

While the expression of quality requirements entails a moderate effort, the actual process of measuring is more difficult and costly. As a consequence, every applicable technique that reduces measurement cost has a great interest when applying quality measure in practice. Here we list a set of techniques we have been used in several evaluations. Their combined applications allows us to make viable the evaluation itself.

- A first technique involves the exploitation of quality requirements: when a specific characteristic is required to be low, it is generally useless to measure it. Since it is possible to trace the results of the measure to the measurement instruments, it is also possible to prune the measurement apparatus in order to reduce costs.

  This pruning technique may be used in a straightforward way when the required quality profile is expressed on the whole application. When using a quality matrix, the pruning technique is still possible, but its effectiveness depends heavily on a good structuring of the software architecture. Namely, if the traceability from functional requirements to architectural components is good, it is easy to identify measures to be taken. On the other hand, when all architectural components are invoked to perform each task, the measurement effort rises abruptly. This points to an additional quality characteristic, and namely the *measurability*, that is the ability of a system to be easily measured and controlled, which is already part of the state of the art of, for instance, electronical engineering. We can think of software systems explicitly designed to be easily measured.

- A second technique is the development of a measurement apparatus which has three different levels of precision (minimal, general and detailed level), as described in the previous paragraph.

At the first level the evaluation requires an effort of the order of magnitude of one working day.

At the second level of assessment the evaluation has costs significantly higher; they are dependent on the size of the application, and may be in the order of magnitude of work months for a large project.

At the third level the evaluation has costs generally not worse than the preceding level, but while in the former case it is possible to reuse existing apparati from the literature, in this case it is necessary to put additional effort in the construction of a specific apparatus. This consideration is the most important motivation for the selection of application classes, in which quality matrices and measurement apparati may be reused, with significant economies of scale and reuse of past experience.

- A third technique is to reduce the man power requested for the evaluation. In order to reach this goal two different strategies could be followed.

  1. Involve as much as possible during the evaluation both the software developer and the user

  2. Use a technique analogous to *the alpha - beta procedure* [7]: first use the cheapest instruments, and stop measuring as soon as further measurements would not change the final result (in term of high, medium, low evaluation).

- Yet another technique for pruning the measurement apparatus, based on the concept of criticality, has been presented in [8].

5. The case study

We applied an evaluation procedure based on quality model, enriched by the techniques and the tools shown above to several products in the fields of scientific computing and of industrial automation. To somewhat clarify the concepts, we discuss a case study, which will be indicated as product X.

Product X is a large industrial automation software product; its size exceeds 500K lines of C code. It will be installed in 100-200 sites and its installed life is expected to be at least 10-20 years. Even if it is not a safety-critical application, failures may cause significant lost profits for the user.

From an organisational viewpoint, product X results from the collaborative efforts of Enel (the Italian National Electricity Board) and of a large supplier. Since maintenance and extensions will be provided by the supplier, a high quality of the product is a desirable goal both for the supplier and for the customer.

The quality evaluation of product X started with a phase of general analysis of the context, which lasted about one month and involved several meetings with representatives of the supplier and of the customer and the analysis of the available documentation.

The following schema (Fig. 6.) explains the subsequent steps of the evaluation:

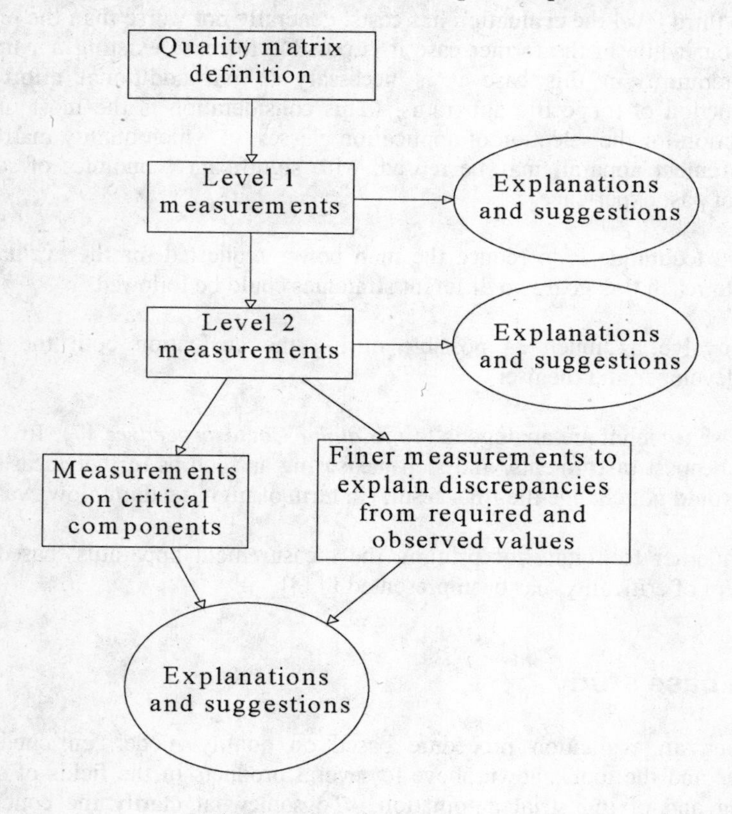

Fig. 6. Steps of the evaluation

A questionnaire was used to elicit from domain experts required quality values for the ISO characteristics. We note that the use of a formal set of questions changed somewhat the judgement of the experts about the required values. For instance, experts informally expressed the belief that usability should have been high; when faced with a series of objective questions about factors implying a high usability, the requirement dropped to a lower value. The value of the questionnaire is to give the expert a structured way for expressing domain knowledge.

Given the size of the product, it was deemed necessary to use a functional decomposition of the product and to express quality requirements on the single functionalities. The decomposition was performed integrating standard textbook knowledge of the domain and the specific functional requirements of product X. Such an integration was deemed necessary because both the supplier and the customer had a deep knowledge of the problem at hand, and tended not to specify the most obvious characteristics of the problem, focusing on what differentiated that problem from similar ones.

Measurements were performed at three increasing levels of assessment; each set of measurements generated some explanations and suggestions. The finer levels were limited to the critical components and to characteristics for which observed values differed from required ones.

The work time of the evaluators may be quantified in three work months. The results were presented to the supplier and to the customer; an example of graphical representation, generated by IQual[1] (and plotted with Microsoft Excel), is depicted in figure 7.

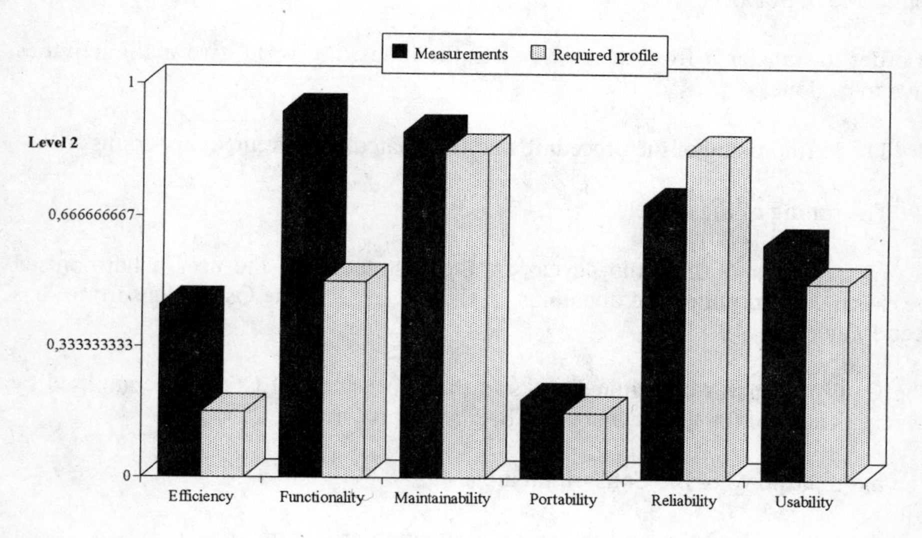

Fig7. IQual: an example of a quality profile

[1]Values are completely fictitious

We gave an explanation about the results; we chose to use for explanation purposes the same model used for the integration process, and to have explanations in form of paths in the model. For instance, if maintainability would be low, it might be traced to a low value of testability, which in turn would point to a low value of traceability (ability to trace code fragments to design entities, and in turn to trace design entities to requirements). Moreover, using a quality matrix, if a single functionality would not pass the quality requirements, we were able to express such detailed explanations as "The sensor diagnostic functionality is not enough maintainable because of low requirement traceability".

We outlined a positive situation with some problems; the supplier confirmed us of the problems (discovered in independent ways) and explained that some corrective actions had been already taken. We understood this point as a positive confirmation of our results. The supplier expressed the intention of implementing further corrective actions using the explanations of the results, and asked for a further evaluation later on.

6. Conclusions and future directions

Enriching the evaluation procedures and related measurement instruments through the techniques described above, we created a measurement procedure really applicable in practice.

In order to transfer it from the research to the industrial world, two main activities have to be done:

- The harmonising of the procedure and the related measurement apparatus
- The tuning of the model

In order to offer both to the developer, the customer and the user a harmonised measurement procedure and apparatus, at the end of 1994 the Qseal Consortium has been founded.

The Consortium was constituted tanks to the effort of CIMECO; it is composed by Cesvit, Etnoteam, Ismes, Tecnopolis CSATA Novus Ortus and IMQ.

The main purposes of the Consortium are:

- to define and publish an harmonised evaluation procedure
- to specialise the measurement apparatus for different kinds of applications
- to offer to the software developer, customer, and user a set of well defined services.

The most reasonable way to tune the model is trying to correlate the a priori (measurements on the product itself) estimates with a posteriori facts (data on the program use, faults, maintenance costs, and so forth); we are sure that in doing that we will incur in sensible difficulties.

In fact, in the literature, all researchers seeking to build meaningful correlations found it necessary to apply an apparatus to a stable environment and/or in a fixed application domain. For instance, [9] analysed a software factory for a bank, while [10], [11] and [12] were focused on a corporate software factory.

Specifically, in our opinion, it does not make sense to compare measurements taken on products developed by organisations at a different maturity level, and with different development processes. For instance, the maintenance effort versus the complexity of the system may have a completely different shape in the case of different organisations, as is shown qualitatively in the following graph (Fig. 8).

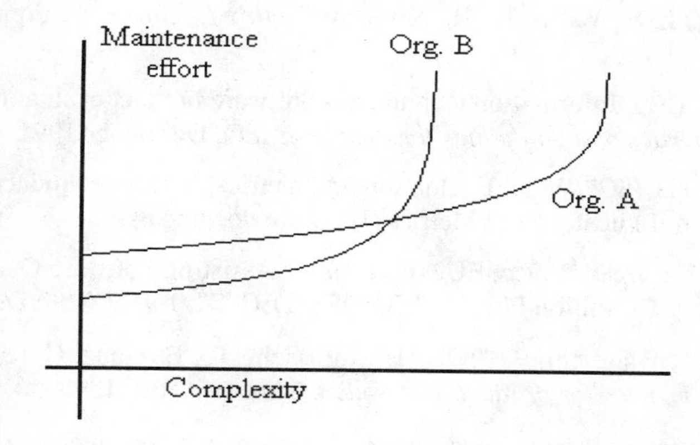

Fig. 8. Maintenance effort versus the complexity of the system

The graph expresses our common-sense belief that more mature organisations (org. A) are able to handle maintenance of complex systems in a different way, and that the area of *manageability,* in which the complexity versus cost is linear is positioned in different parts of the complexity line. As a consequence, the same complexity measure is not able to estimate the maintainability of a product in both cases.

On the other hand, if we compare two software systems of different nature with the same apparatus, it is evident that a posteriori data in not directly correlated with a priori measurements. For instance, scientific computation codes have cyclomatic numbers consistently higher than process-control codes. This means that, in order to find correlations, it is necessary to collect data in the same application domain.

The motivations for trying to correlate a priori estimates with a posteriori facts should be self evident: it would be possible, for instance, to estimate maintenance costs, failure rate with relatively simple and cheap measurements. Moreover, some sort of explanation of result would be available, making it possible to retroact early on the process with specific corrective actions.

A future goal of our research is to build a data base of measurement and observed results, in order to calibrate a small set of quality models. Statistical techniques would be used to tune model parameters to fit observed data. However, in order to make statistical techniques more effective, it will be necessary to reduce the complexity of the available models.

7. References

1. Bowen, T., P., Wigle, B.,Tsai, J., T., *Specification of Software Quality*, RADC-TR-85-37, 1985

2. Deutsch, M., S., Willis, R., R., *Software Quality Engineering*, Prentice Hall, 1988

3. ISO/IEC 9126, Information technology- Software product evaluation- *Quality characteristics and guidelines for their use*, ISO, December 1991

4. ISO/IEC JTC/SC7/WG6, Evaluation and metrics, Software product Evaluation, Indicators and Metrics, Working documents.

5. TASQUE- Eureka Project EU240 "Tool for Assisting Software Quality Evaluation: Definition Phase", TSQ/DEF/TEC-T&/10/1.2/0290/DEL

6. Pina, D., Salvaneschi, P., "IQUAL", In: Bache, R., Bazzana, G. (eds), *Software metrics for product assessment,* Mc Graw Hill, 1993

7. Nils J. Nilsson, *Principles of Artificial Intelligence,* Springer-Verlag, 1980

8. Pina, D., Salvaneschi, P., Boninsegna, A., Zambetti, R., "Tailoring V&V Plans through Expert Systems technology", *Third European Conf. on Software Quality,* Madrid, 1992

9. Zage, W., M., Zage, D., M., "Evaluating Design Metrics on Large-Scale Software", *IEEE Software*, July 1988.

10. Koga, K., "Software reliability design method in Hitachi", in: *3rd European Conference on Software Engineering*, Madrid, 1992.

11. Sheldon, Kai, Tansworthe, Yu, Buttscheider, Everett, "Reliability Measurement: from Theory to Practice", *IEEE Software*, July 1992.

12. Shen, V., Y., Yu, T., J., Thebaut, S., Paulsen, L., "Identifying Error Prone software - An Empirical Study", IEEE Transactions on Software Engineering, Vol SE-11, No. 4, April 1985.

Quantifying the Benefits of Software Testing: an Experience Report from the GSM Application Domain

Gualtiero Bazzana[1], Roberto Delmiglio[2],
Aldo Lora[2], Olivia Balestrini[1], Silvana Finetti[1]
[1]Etnoteam S.p.A, Via A. Bono Cairoli, 34, 20127 Milano (ITALY) - gbazzana@etnoteam.it
[2]Siemens Telecomunicazioni Italia, SS Padana Superiore, Cassina de Pecchi (ITALY)

Abstract. This paper aims at describing the quantitative results gained from a systematic integration test campaign undergone at Siemens Telecomunicazioni Italia on parts of a GSM (Global System for Mobile Communications) Phase 2 system. In particular, the following topics are dealt with: basics of the Siemens development process, with emphasis on testing issues; a high level description of the system under test; the actions taken in order to improve the effectiveness of software testing practices; the methods and techniques applied during the project; the measurement system that was set-up in order to track the effectiveness of the improvements; the quantitative results collected, with associated interpretations; the future goals, in the light of continuous improvement. The results clearly show a positive impact of systematic testing on timeliness, reliability and documentation, underlining also the importance of anticipating error detection as much as possible in the development cycle and the need for tools supporting the various activities. The data collected provide also models and baselines onto which it will be possible to base estimates for new projects.

1 The Software Producing Unit and Its Development Process

The experience described in this report has been matured at Siemens Telecomunicazioni Italia (STI), in the context of a project committed to the development of a telecom system for mobile phone handling, in accordance to the GSM International Standard, phase 2.

The development process at STI - Mobile Phone Division is ISO 9001 conformant and adopts a classic waterfall model, characterised by system analysis, software specifications, architectural design, coding and debugging, integration, qualification and acceptance with the customer. As far as telecommunication systems are concerned, it is very seldom the case in which development starts from scratch. For this reason, the development normally feature-driven, that is to say that the afore mentioned phases must be performed for each additional feature to be developed onto an existing version of the system. Moreover, the peculiarities of the products developed (big size and long life) have also brought to well founded and applied practices as far as supporting activities are concerned (e.g.: configuration management, document control, etc.). The paper focuses on the quantitative benefits gained from a systematic approach to integration testing activities, by which it is meant the validation step performed by the project before delivery to system test.

2 The system under test

GSM is the European standard for the European 900 MHz mobile telecommunication Digital Cellular Systems [1], allowing telephony services through mobile phones. Mobile telecommunications is not a very recent technology, but it is a rapidly evolving one. The main differences with wireline telecommunication access consist in:

- mobility management: as a consequence of the fact that subscribers can continuously change their point of access to the network, routing of calls becomes a major issue and involves new concepts like: location management, handover (automatic transfer of a call in progress from one cell to another without inconvenience) and roaming (free circulation of mobile stations across networks handled by different operators)
- radio resource management: the link between the subscribers and the fixed infrastructure is not permanent and wave propagation limits and spectrum scarcity have to be taken into account.

From the architectural point of view, GSM is a complex object, since it has to deal with multi-services and with the difficulties coming from cellular networks. Looking at the system from the outside, GSM is in direct contact with users, with other telecommunications networks and with the personnel of the service providers. The internal GSM architecture distinguishes two parts: the BSS (the Base Station Sub-System, that is in charge of providing and managing transmission paths) and the NSS (Network and Switching Sub-System, that is in charge of managing the communications). Getting into details of the BSS, we can find three types of machines: a transmission equipment (the BTS - Base Transceiver Station), a managing equipment (BSC - Base Station Controller) and a speech encoding/ decoding equipment (TRAU - Transcoder and Rate Adapter Unit, that performs also rate adaptation in case of data and, though considered a sub-part of the BTS, is often sited away as a stand-alone equipment).
The goal of the project was to develop the BSC and TRAU Network Elements, together with a Local Terminal (named: LMT) for managing the equipments. A BSC (that was the biggest effort of the project) can be considered a small switch composed by several boards, and providing a wide range of features including call processing (management of channels on the radio interface, handover, mobile originated and terminated call, location update, call assignment, short message services, encryption, frequency hopping, etc.) and operation and maintenance (download and alignment of Network Elements, management of objects and associated status, performance measurement, redundancy management, detection and correction of faults, etc.)
The testing of such systems is not a trivial task, considering that, besides the usual problem posed by big systems [6], tye following aspects have to be taken into account: the target systems are quite complex; the development is sub-divided among teams that are geographically distributed (multi-site development); the configurability of the systems, together with the presence of standardised interfaces,

cause an exponential growth of situations to be considered; the test beds (environment and tools) need to be prepared ad-hoc; the wide range of layers to validate (firmware, operating systems, transmission protocols, application software dealing with call processing, application software dealing with operation and maintenance, etc.) requires utterly different approaches in test design and execution; the many to many relationship among physical objects (processors and executable processes) and features imposes an accurate planning of deliveries and synchronisation points across the various development teams.

3 Characteristics of the project

The software development project to which the experience applies, is characterised by:

- Time scale for software development and integration testing: 12 months (interspersed with several sub-releases, known as "builds");
- Product size: approximately 500 KLOC (Kilo Lines of Code) mainly in C language;
- Re-use from previous products: approximately 50%;
- Project staffing: approximately 60 Full Time Equivalent (including developers, testers, supervisors and support staff but excluding: configuration management, system test and project management staff).

4 Goals of the testing improvement initiative

For the afore described project it was decided to strengthen the testing activities, by means of the creation of a testing team within the development structure (known as: TDS - Test Design and Support Team) having the following goals:

- to devise technical solutions allowing testing execution on host as soon as possible, without waiting for the availability of target prototypes; this was a major change to the development process, since it involved an additional activity (namely: host testing) that had to be supported from the organisational, methodological and technical points of view;
- to speed up integration of sub- systems and Network Elements;
- to ensure that most failures are captured before the Network Elements are shipped to system test;
- to validate non-regression of existing features;
- to promote testing automation;
- to apply a disciplined approach to testing, in accordance with Siemens Quality Management System guidelines and providing a support to developers in their activities.

The resulting life-cycle is detailed in Fig.1, where acronyms stand for the various document classes produced as outputs of the life-cycle phases.

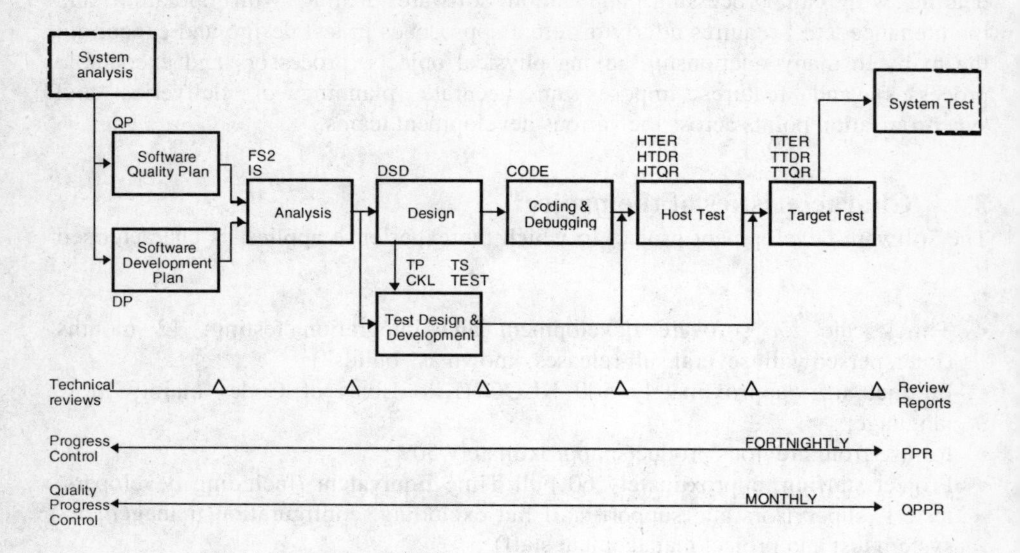

Fig. 1. The development life-cycle adopted

5 The testing strategy: methodology, techniques and tools

The integration testing approach (described in the "Entity Integration Test Plan" of the project) is based on a V-cycle, with test design activities done in parallel with code development and debugging; test design is based on a functional black-box approach focusing, for test derivation, on analysis and design documents (known as: Functional Specifications level 2 and DSD - Delta Design Documents). The test design phases are shown in Fig. 2, with evidence of associated documents.

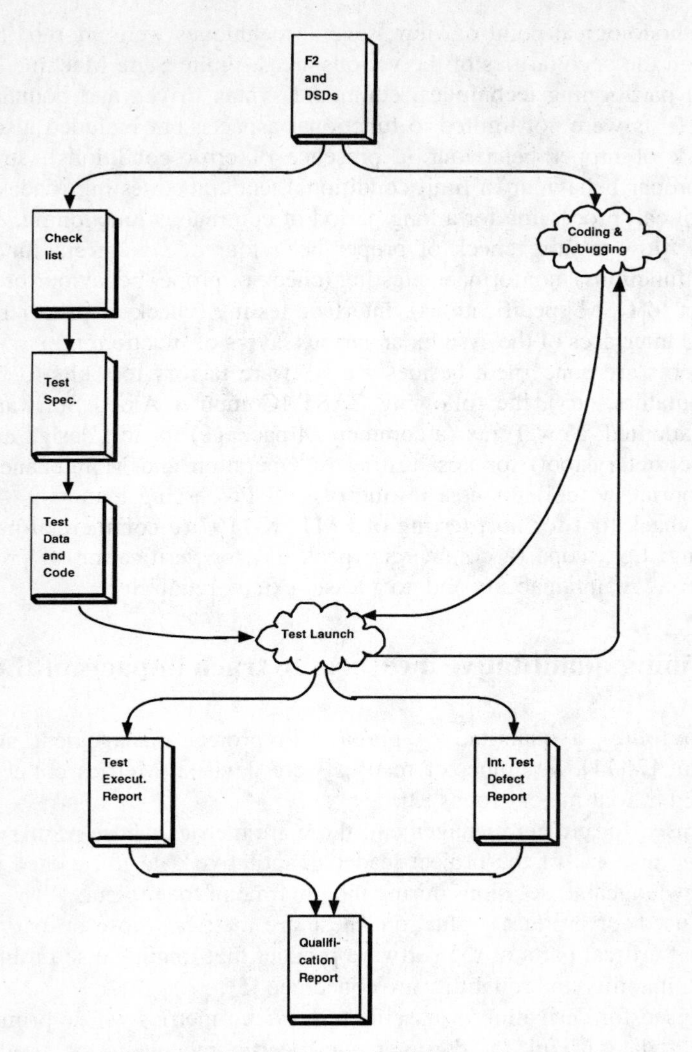

Fig. 2. The testing methodology adopted

Several levels of integration testing were designed and executed, among which:

- Host test, in terms of systematic test, in a simulated environment of: single tasks; tasks in a processor; integration of tasks of different processors in the same functional area
- Target test, in terms of systematic test, in the final environment of: integration of tasks of different processors in the same functional area; integration of tasks of different processors of different areas contributing to a feature; overall features; integration of features in the same Network Element; work flow among Network elements.

From a methodological point of view, several techniques were adopted test design, depending on the peculiarities of the various areas: Finite State Machine techniques; input space partitioning techniques; command syntax driven and boundary checks techniques.Tests were not limited to functional aspects, but included also: negative cases (check of proper behaviour in presence of error conditions), stress testing (check of proper behaviour in limit conditions), endurance testing (check of proper behaviour of call processing for a long period of continuous functioning, e.g. several days), work flow testing (check of proper behaviour of concurrent functioning of interleaved functions), conformance testing (check of proper behaviour of the system with respect to GSM specifications), interface testing (check of proper handling of the standard interfaces of the system, at various layers of functioning).
As far as tools are concerned, besides the software factory tool-kits (debuggers, In Circuit Emulators, etc.), the following CAST (Computer Aided Software Testing) tools were adopted: New Tefax (a commercial package) for test design and control; Flute (a proprietary tool) for host testing of Operation and Maintenance features; RTA (a proprietary tool) for host testing of Call Processing features; a share-ware capture-playback tool for host testing of LMT; K-1197 (a commercial package) for target testing; Logiscope (a commercial package) for verification of programming rules impacting maintainability and, to a lesser extent, reliability.

6 Defining quantitative measures to track impacts of the test campaign

In order to follow a quantitative approach in project management and process improvement [7][11], a number of metrics were devised. Metrics collected can be distinguished in four major groups [2]:

- metrics used for project management: these are metrics whose primary goal is to put at the disposal of the project leader quantitative data to be used in order to take knowledgeable decisions during the life time of the project;
- metrics used for product evaluation: these are metrics whose primary goal is to single out critical parts of the software product that might cause problems as far as maintainability and reliability are concerned [3];
- metrics used for derivation of baselines: these are metrics whose primary goal is to collect data useful for deriving site-specific models to be used in future projects as estimate rules;
- metrics used for evaluation of process improvement effectiveness: these are metrics whose primary goal is to quantitatively derive a judgement on the usefulness of the initiatives, in order to decide whether to adopt them continuously or to reject them.

It has to be underlined that the testing metrics were part of a bigger management-by-metrics initiative embracing also development activities and thus topics like: development productivity, project timeliness, etc. In the following the various metrics are listed in accordance with the classification presented earlier.

6.1 Metrics for project management

Such metrics were used for deciding project re-scheduling and readiness for delivery.

Goal	Adopted metric	Source of data
Progress tracking	Number of tests executed vs. planned	Test planning +
	Status of testing documents	Design document
	Failure distribution vs. severity and status	
	Time for fault removal	
Reliability	Fault density vs. Size of software	Anomaly Journal +
	Fault density vs. Tests executed	Sw size +
	Fault Rate	Tests executed

Tab. 1. Metrics for project management

Readiness for delivery was decided on the basis of evidence of the following quantitative conditions:

- tests designed =100%
- functional tests executed = 100%
- functional tests successful > 95%^
- non regression tests executed = 100%
- non-regression tests successful = 100%
- call completion rate > 99,5%
- open fault report density < 0,5 Faults/ KLOC

6.2 Metrics for product evaluation

As far as product evaluation is concerned, a number of metrics were defined basing on source code structural complexity theory [4], as described in the following. Such metrics were used as non-mandatory indications on which parts to re-engineer when new features had to be developed.

Goal (Sub-char of ISO 9126 [10])	Metric	Lower Bound	Upper Bound
Analysability	Statement Size	3	7
	Comment density	0,2	0,65
Testability	Cyclomatic number	1	10
Stability	Control Density	0	0,2
	Number of Statements	5	100
Changeability	Nesting Levels	1	6
	Max. Distance	0	25

Tab. 2. Metrics for product evaluation

The metrics (calculated for each function contributing to the source code of the product) were integrated using weighted composition algorithms in order to derive a single "maintainability index" (spanning in the range from 0 to 1) at various levels of granularity (function, process, functional area, processor, network element, product). Having defined as goal a target level no less than 0.7, functions with a maintainability index below 0.6 were proposed for reverse engineering.

6.3 Metrics for derivation of baselines

The derivation of baselines is a fundamental step toward the derivation of predictive models based on quantitative experience. Such metrics are deemed as less important than others at project level, since they are more useful for future projects than for the current one (that experiences overhead for data collection). In this situation the presence of several builds allowed the usage of predictive models already during the life-time of the project.

Goal	Adopted metric	Source of data
Test design estimate	Test design productivity	Test effort data
Test execution estimate	Test execution productivity	Test effort data +
	Usage of test environment	Tests executed
Project oversight	Distribution of effort across phases	Effort data

Tab. 3. Metrics for derivation of baselines

Metrics were collected at quite a fine granularity level, in order to have at disposal of supervisors specific values for the various areas (that deal with software of varying nature). In order to overcome possible resistance from the staff, it was made clear that the metrics were collected for the sake of process assessment and improvement through management-by-metrics and that no usage of the metrics was foreseen or intended towards judging the performance of individuals or groups.

6.4 Metrics for evaluation of the improvement initiative

These metrics were defined in order to track the effectiveness of the process improvement initiatives. As such, the metrics were used to evaluate the usefulness of the test campaign and of the associated methods, techniques and tools. Such metrics were constantly kept under control alongside the life of the project, in order to single out improvement opportunities.

Goal	Adopted metric	Source of data
Impact on reliability	Test effectiveness	Anomaly Journal
Impact on documentation	Number of tests and functionalities	Test documents +
	Number of tests vs. product size	Sw size
Impact on timeliness	Test timeliness	Test plans
	Delivery timeliness	
Impact on productivity	Overhead of test campaign	Test effort data
	Distribution of effort across test design and support activities	
Effectiveness of methods	Failure distribution vs. areas	Anomaly Journal
Effectiveness of tools	Failure distribution vs. environment	Anomaly Journal

Tab. 4. Metrics for validation of best practices

7 Quantitative progress control: life on the critical path

Testing is by its own nature an activity that has to deal with stringent deadlines and timeliness constraint. For the project in question these aspects were magnified, considering that an additional testing level had been introduced (off-line testing) and that the acceptance of deliveries was subjected to the evidence of reaching a satisfactory testing level. The key factors for succeeding in keeping the deadlines were two: the adoption of host testing and the accurate planning of test sessions at the target.

The former aspect resulted in fact not in a delay but in a strong contribution to timeliness achievements; in fact the availability of an off-line environment allowed designers to remove most errors before integration at target took place, thus speeding substantially the overall system integration, that had to concentrate on interface faults rather than on logic and trivial faults that had already been removed. Tab.5 shows the number of tests executed on host and target for various areas alongside the various builds; the number of executed tests is big, especially considering the timing requirements for testing (normally the testing of each sub-release had to be performed in less than two months of elapsed time).

	Build 1		Build 2		Build 3		Build 4		Build 5		TOTAL		
	H	T	H	T	H	T	H	T	H	T	H	T	H+T
A1	179	184	28	96	723	510	29	20	68	105	1027	915	1942
A2	77	30	190	177		220		66	70	83	337	576	913
A3			798	511	442	296	74	140	0	15	1314	962	2276
A4	227	61	216	210	227	111	41	220	0	47	711	649	1360
A5			11	33		66		10	7	90	18	199	217
A6			585	470	84	85	179	262	0	54	848	871	1719
A7			97		499						596	0	596
A8	151	53	0	65	490	155	57	376	82	226	780	875	1655
A9	1454	37	5536	37	18698	63	1226	37	470	0	27384	174	27558
A10		12		55		50		201			0	318	318
Tot	2088	377	7461	1654	21163	1556	1606	1332	697	620	33015	5539	38554

Tab. 5. Tests executed

The latter aspect (accurate planning of target testing) is typical of embedded systems whose complexity and cost cause a shortage of test beds with respect to the potential needs of the project. In order to make the most effective usage of resource available (machines and time) the following working scheme was adopted:

- testing activities for the specific build were broken down at a very fine granular ity level in a set of work packages;
- for each testing activities for the specific build were broken down at a very fine granular ity level in a set of work packages;
- for each work package the following information were collected: number of tests to be executed, other work packages whose execution constituted a pre-condition, requirements in terms of test beds;
- starting from the afore mentioned information, a PERT was prepared linking together all work packages and detailing the number of tests to be executed;
- the PERT was optimised in order to succeed in executing all tests of all work packages before the delivery deadline, allowing also for a final non-regression and for periodic maintenance activities on the test beds;
- starting from the optimised PERT, each week a detailed allocation of test beds was produced;
- the PERT was then used for progress tracking and control.

Progress tracking and control was based on Progress Trend Analysis, that is to say by means of the comparison of expected test execution trends versus actual ones. Such analysis proved extremely effective to detect critical points early and thus to adopt corrective and preventive actions. For instance, just looking at the expected growth patterns it was possible to easy recognise the critical areas (e.g. those with a "stairway" curve or with most tests concentrated in the last time period) with respect to the standard areas (characterised by a S-shaped rate growth pattern).

8 Product evaluation: fighting against entropy

As far as product evaluation is concerned, the adoption of static analysis techniques allowed designers to single out critical functions and to master the complexity growth of source code. The analysis of data showed that:

- areas where a major reverse engineering was made had a substantial growth of the maintainability index;
- areas where specific structural changes were made in accordance with suggestions from static analysis metrics, also showed a significant bettering;
- areas where no specific actions were adopted succeeded in keeping to a minimum the impact of additional features within existing source code.

In conclusion it is possible to say that the adoption of static analysis techniques was positive, since designers focused their reverse engineering efforts on error-prone modules, in accordance with a Pareto strategy (that suggests, in order to maximise the return on investment from process improvement, to focus on the 20% of the aspect that is likely to cause the 80% of the troubles).

9 Derivation of baselines: from control to oversight

During the project, emphasis was given in the collection of effort data at appropriate granularity levels in order to fine tune the estimate models in use.

As far as test design productivity is concerned, the average value stabilised on 7 tests fully specified per person-day, but with significant variations across areas. The early availability of such baseline allowed to significantly better the estimates on test design activity, as reported in Tab. 6 that shows the variations in estimate errors, depending on the techniques used for doing the estimate.

As far as test execution productivity is concerned, Fig. 3 shows the test execution rates exhibited by various areas (having varying complexity and different testing tools). The early availability of such baselines guaranteed the timeliness of test activities

Estimation technique	% Estimate error	Error range
Qualitative analysis based on personal experience	92,6%	min = 13% max. = 584%
Qualitative analysis, with the support of domain experts	81,1%	min = 31,9% max. = 232,6%
Quantitative estimation reviewed after 10% work completion, with usage of project baselines	18%	min = 12,4% max. = 31,6%
Quantitative estimation done reviewed after 10% work completion, with usage of application specific baselines	12,1%	min = 0% max. = 30%

Tab. 6. Impact of baselines on estimate accurateness

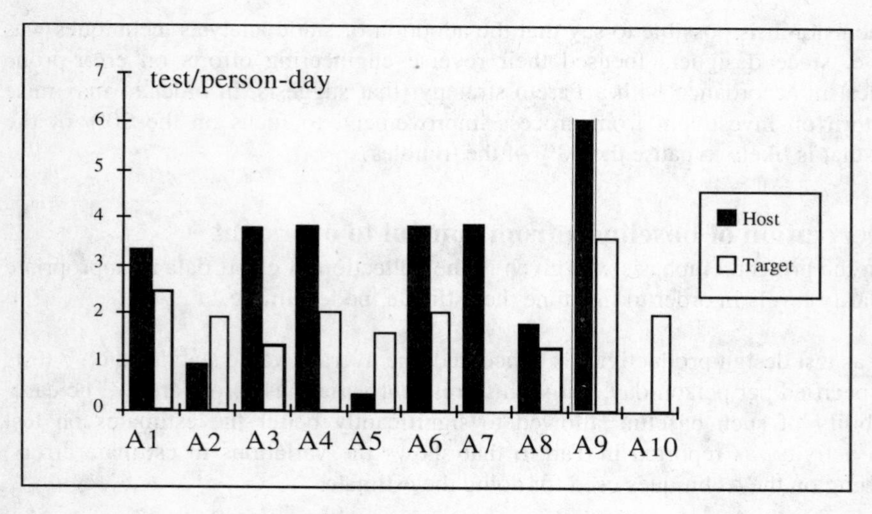

Fig. 3. Test execution metrics

On a more general perspective, the distribution of effort with respect to the various phases of the development life cycle was calculated, for the sake of future projects. The results (aligned at the end of development and thus scarcely taking into account maintenance activities on defect reports found by system test) are reported in Tab. 7, compared with the estimates done at the very beginning of the project based on company-wide reference model: the existence of several differences underlines the importance of site specific models. As far as testing is concerned, it is possible to note that the adoption of the process improvement initiatives brought to a general reduction of the weight of test execution phases: this is a very good result and clearly shows a positive return on investment.

Phase	Estimated Impact	Current Impact
Analysis and Design	21%	30%
Coding and Debugging	16%	24%
Execution of host tests	16%	12,7%
Execution of target tests	27%	17%
Maintenance	-	1%
Co-ordination	10%	4%
Review	-	0,3%
Support to external groups	-	3%
Test design	10%	8%

Tab. 7. Effort distribution across phases

10 Process improvement: to test or not to test?

The success of process improvement in testing can be monitored through a number of questions, for instance:

- which percentage of failures where discovered by the integration phase (test effectiveness indicator)?
- what kind of information will be available for future projects (documentation indicator)?
- what is the impact as far as delivery deadlines are concerned (timeliness indicator)?
- which is the overhead of testing activities (cost indicator)?
- considering the specific technical aspects of the project (focus on host testing), what has been the effectiveness of introduced techniques and tools?

In the following we will try to give quantitative hints on all the afore mentioned issues. As far as testing effectiveness is concerned, the integration test activities discovered about 2000 failures, yielding a testing effectiveness (calculated as the ratio between failures discovered in integration test and those discovered in integration + system test) above 80%. In particular the testing activities were able to halve the fault density from a testing phase to the following one, obtaining a result comparable to the experiences reported by the SEL [5]. As far as documentation issues are concerned, Tab.8 shows the number of tests documented (in terms of: test purpose, preconditions, requirements for test bed, input data, expected results) and put under configuration management. Tests have been packaged in several test specification documents and will be used for non-regression of future releases.

Area	A1	A2	A3	A4	A5	A6	A7	A8	A9	A10	TOT
Numb of Tests	951	703	1323	537	333	944	496	1117	337	415	7156

Tab. 8. Number of tests designed

As far as timeliness is concerned, test activities proved to contribute positively, as detailed in the chapter about "Project Management". As far as cost issues are concerned, the overhead of test design and support activities accounted for 8% of the overall software development. Fig. 4 shows also the distribution of effort across the various test design and support activities. As it is possible to see, the overhead of the management-by-metrics program accounts for about 9% of the TDS effort (summing measures collection and static analysis), that is to say less than 1% of the overall software development effort.

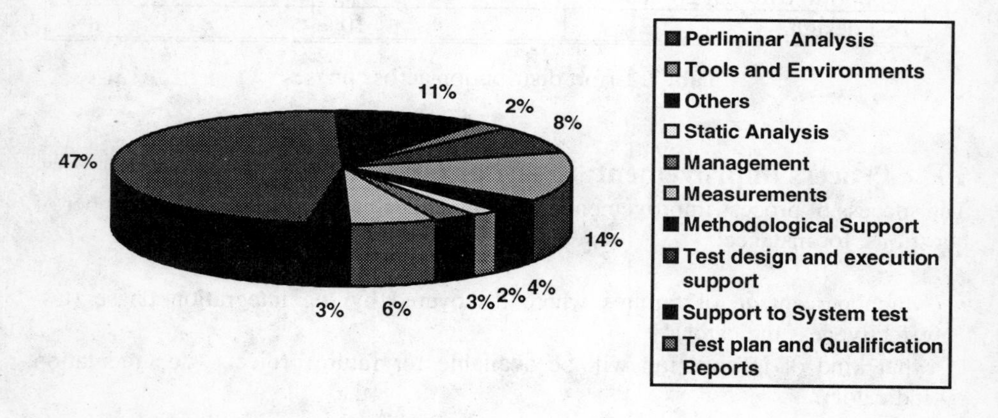

Fig. 4. Distribution of test design and support effort

Last but not least, can we say anything about the effectiveness of off-line testing? Indeed, we can say that the introduction of such aspect resulted in significant improvements, considering that:

- most failures were captured off-line (62%, see Fig. 5); moreover, the analysis of failures found on the target environment showed that only 17% of those were detectable on host: most were in fact by their nature dependent on the physical configuration of test beds.

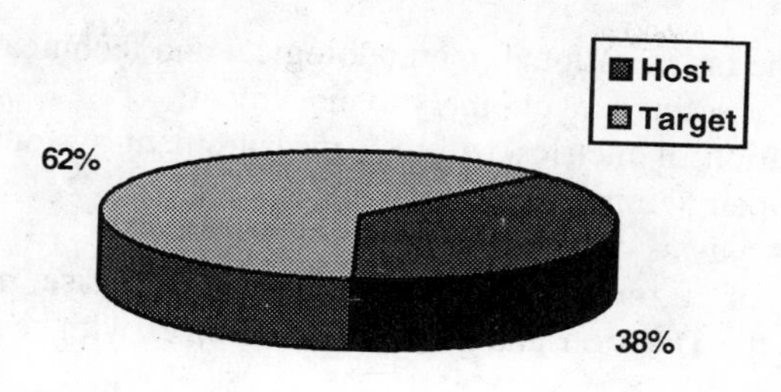

Fig. 5. Distribution of failures vs. environment

- as previously shown in Fig. 3, host testing yielded a double productivity with respect to target testing (that was in any case more efficient than initially estimated, thanks to the fact that most faults had already been removed from builds);
- the test - fix - validate cycle was much quicker on host than on target, as reported in Tab. 9, that shows an exponential growth of fixing effort with respect to various testing phases, as taught by most software engineering text-books!

Testing phase	Integration - host	Integration- target	System test
Effort in bug-fixing (person-hours)	1,34	3,73	21,55

Tab. 9. Effort in fault removal

11 Final considerations

To sum up, we consider that the process improvement initiatives that embraced the testing activities proved to be fruitful with respect to timeliness, reliability and documentation of the specific project. Moreover, the availability of quantitative data should improve the capability of process control of other projects.
For this reasons, systematic testing and process improvement will continue in the next release of the product and will also be extended and strengthened, by means of:

- a focused training initiative;
- the creation of a small Software Engineering Process Group;
- the enhancement of the testing environment (both by extending currently adopted tools and by developing new ones);
- the widespread usage of static analysis tools, that will be used also for deriving checks to be done during debugging, in order to ensure an appropriate topological coverage;

- the development of tools calculating the size of products (both in terms of LOC and in terms of lines added, modified, removed) in adherence to company guidelines;
- the tuning of the organisational, methodological and technical aspects concerning test execution reporting and anomaly management;
- the standardisation of metrics within a measurement system adopted by all the projects developed within the Division;
- a Root Cause Analysis of failures;
- the execution of a formal process assessment exercise, in order to get the positioning of the Division and the specific project with respect to the maturity level grading [8] [9];
- the extension of the test methodology in order to accommodate feature-driven test design.

ACKNOWLEDGEMENTS

Our thanks go to G. Cecchetto, E. Pietralunga and G. Vulpetti who supported and sponsored the initiative; we also have to thank B. Marelli, O. Fouillouze and M. Maiocchi for overall supervision of the activities. The improvement has been supported by the whole project staff, headed by the development supervisors: P. Bettoni, B. Ferri, A. Manini, L. Travaglini. The collection of quantitative measures was supported by the Configuration Management Team, headed by S. Scotto di Vettimo, and by the Project Management Team, especially: D. Gandini and S. Zanella. Last but not least, a particular mention has to be devoted to the TDS Team, that performed most of the work described: A. Antonelli, A. Armelloni, L. Barbieri, S. Benetazzo, G. Bollani, P. Bottazzi, R. Brigliadori, O. Cantoni, L. Careddu, A. Clima, E. Colombo, P. Consolaro, L. Copercini, M. Crubellati, M.G. Ottolina, G. Perricone, F. Rossi, G. Savino, R. Tarquini, S. Viganò.

REFERENCES

1. Mouly, M. Pautet, “The GSM System for Mobile Communications”, Europe Media Publications, 1993
2. Grady, “Successfully applying software metrics”, IEEE Software, Vol.27, No.9, September 1994
3. Bache, G. Bazzana, “Software metrics for product assessment”, Mc Graw Hill, London, 1994
4. McCabe, “A complexity measure”, IEEE Transactions on Software Engineering, 1976
5. SEL, NASA Goddard Space Flight Center, “Software Engineering Laboratory Relationships, models and management rules”, SEL-91-001, February 1991
6. Marks, “Testing very big systems’, McGraw-Hill, 1992
7. Bazzana, P. Caliman, D. Gandini, R. Lancellotti, P. Marino, “Software management by metrics: practical experiences in Italy”, 10th CSR Workshop, Amsterdam, October 1993

8. Weber, M.C. Paulk, C.J. Wise, J.V. Withey, "Key practices of the Capability Maturity Model", SEI, Carneige Mellon University, Pittsburgh, 1991, CMU/SEI-91-TR-24, ADA240604
9. Members of the BOOTSTRAP project team, "BOOTSTRAP: Europe's Assessment Method", IEEE Software, May 1993
10. International Organization for Standardization, "Information technology - Software product evaluation - Quality characteristics and guide lines for their use", ISO/IEC IS 9126, December 1991
11. Moeller, D. Paulish, "Software Metrics: a practitionar's approach to improved software develoment", Chapman & Hall, 1992

Benefits of using model-based testing tools

Giorgio Bruno[1],Mauro Varani[2], Valter Vico[3], Chris Offerman[4]

[1]Dip. Automatica e Informatica, Politecnico di Torino
corso Duca degli Abruzzi 24, 10129 Torino, Italy
Email Bruno@polito.it
[2]ARTIS S.r.l.
corso Cairoli 8, 10125 Torino, Italy
Email ARTIS@ALPcom.it
[3]FIAT AUTO MAINS S.r.l.
via Issiglio 63/A, 10141 Torino, Italy
[4]CAP VOLMAC
Daltonlaan 300, 3500 GN Utrecht, The Netherlands
Email cv2970@inetgate.capvolmac.nl

Abstract. This paper presents a methodology based on modeling that proved to help in developing cost-effective system testing techniques for complex discrete-event systems, such as real-time control systems and supervisors. The main point is to build an executable model of the application's environment, which we call an emulator, for system testing purposes. The emulator allows the testing team to validate the application in realistic workload conditions which can be produced only if the environment is represented by a state-based model able to respond to the application's commands as well as to send unsolicited events. A fundamental role is played by the particular modeling approach used which in this case is based on the Protob language and its support toolset. Three experiences in the area of manufacturing applications are illustrated and their results discussed.

1 Introduction

Testing is the last and most crucial step in the quality chain of the software development process. The importance of testing is rooted in the still very high share of costs that quality assurance procedures and error fixing hold in the budget of software projects.

Reducing development costs very often represents the key to be competitive. The cost of a software project is not only related to the development effort but also to the quality of the process. Sometimes it happens that the lower is the effort spent in the development, the worse is product's quality, but the opposite is not always true. Nevertheless in order to be competitive it is required both to reduce the development time and to improve quality.

Being able to reveal defects earlier reduces the effort needed to fix them (figure 1). Within quality assurance, testing is typically regarded as a last-phase remedy for avoiding uncomfortable outcomes, not as a precaution.

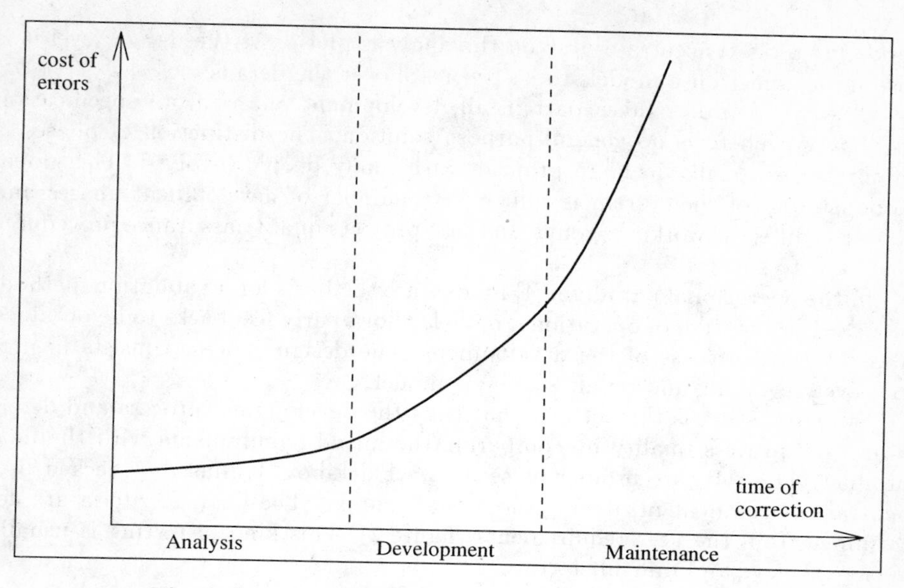

Fig. 1. Late correction of errors is extremely expensive

A typical problem is the availability of adequate testing environments. Testing a single separate module is usually performed by connecting it to some kind of terminator able to feed inputs and check results. Integration testing requires more sophisticated tools to manage the combination of several modules developed in the project and also of external software. During the system testing phase the main concern regards the realism of the tests. Quite often the software cannot be run in the target environment because it is not yet available or it is too expensive or risky to use it. In these cases the problems usually concern how to generate an equivalent set of testing conditions.

We especially focus on complex software systems such as real-time control systems and supervisors. They are an important class of systems as they are generally very expensive to develop, they require high reliability and are very difficult to test exhaustively.

In this paper we present a methodology based on modeling that proved to help in developing cost-effective testing techniques. We report some real applications of it demonstrating how the modeling paradigm can bring valuable benefits to the software development process. We will explain how models improve the understanding of problems by means of a clear definition of systems and processes.

Models are a representation of the system in a form that is easy to understand. Accurate and rigorous models are an excellent means for specifying and documenting a software system. It is quite common, especially in large projects, to use models in the analysis and specification of software systems. Many methods and tools are available. Modeling helps being clear and communicative. Most modeling tools concentrate only on the analysis phase. At the design level it is also important to use models to keep control over the details.

Quality assurance takes part in all development phases from specification to delivery. There is no general-purpose solution. The distinction of phases is important especially in large projects with many people involved. The global responsibility of the system is split over a number of development phases and concerns different working groups. In these projects quality assurance procedures are a key to success.

In the operational paradigm [1] models are the basis for a validation methodology. The execution of operational models allows early feedbacks to be obtained about the correctness of the development. The design process consists in progressively evolving and refining a single model.

System testing is the activity that ends the development process and determines the process quality by comparing the initial requirements with the final product. The software product is seen as a black-box. It must be checked in a neutral environment mirrowing the actual context. The testing criteria are determined from the user requirements (figure 2). This kind of testing is usually very expensive and difficult to run.

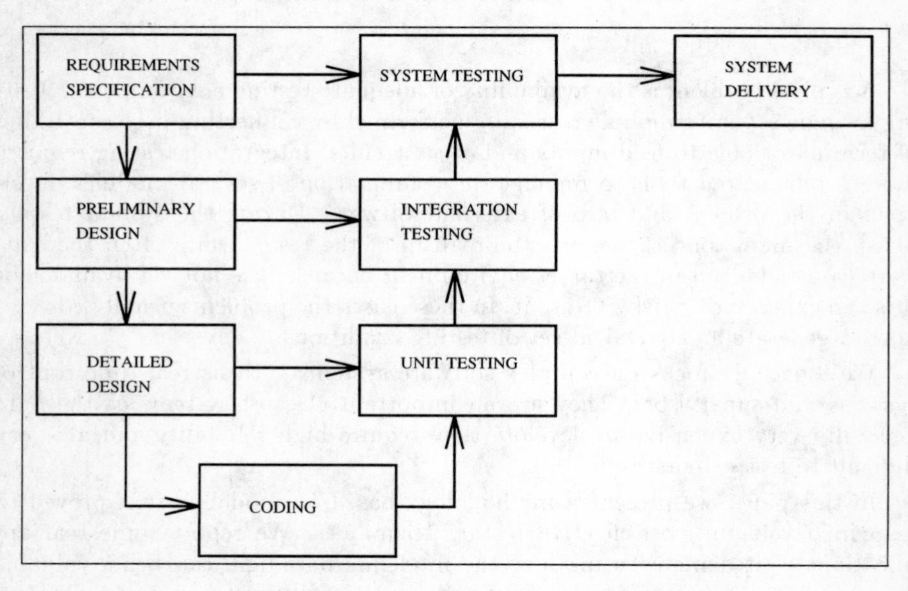

Fig. 2. The testing in the software life cycle

Technology may reduce such costs. But which are the benefits of improving the quality of system testing? In general we expect saved-costs to depend on the consequences of potential errors. It is very difficult to estimate the potential cost of errors. We may at most calculate costs a posteriori. Although it is uncomfortable to admit, it is still everybody's experience that the most serious errors are still found precisely during system testing. They are usually the most subtle, difficult to test and therefore critical.

2 Requirements for system-testing tools

Validation means checking whether the software product is compliant with the requirements. This definition has a legal flavour which is in fact a very important aspect to be considered when dealing with contracts and responsibility. But in this paper we prefer to consider the task of validating software products in a wider context, perhaps more end-user oriented: this activity aims at ensuring that the delivered system will be able to operate correctly in the target environment. The difference roots in the emphasis placed on the realism of the test rather than on the compliance to a contract. Nevertheless the two viewpoints are more complementary than contrasting.

Validation procedures take place at the end of the development process when the software system is integrated and adequately stable. The goal is to evaluate the correctness and reliability of the system and to measure its performance. Validation techniques focus on the definition of test cases (on the basis of user requirements), on the generation of such test cases (this involves providing suitable communication mechanisms between the testing environment and the software product to be tested) and on the management of test cases (including the analysis of the product behaviour).

The most complete and effective technique for testing a software system is to embed it into an artificial environment which reproduces - emulates - the interactions between the software system and the outside world (figure 3).

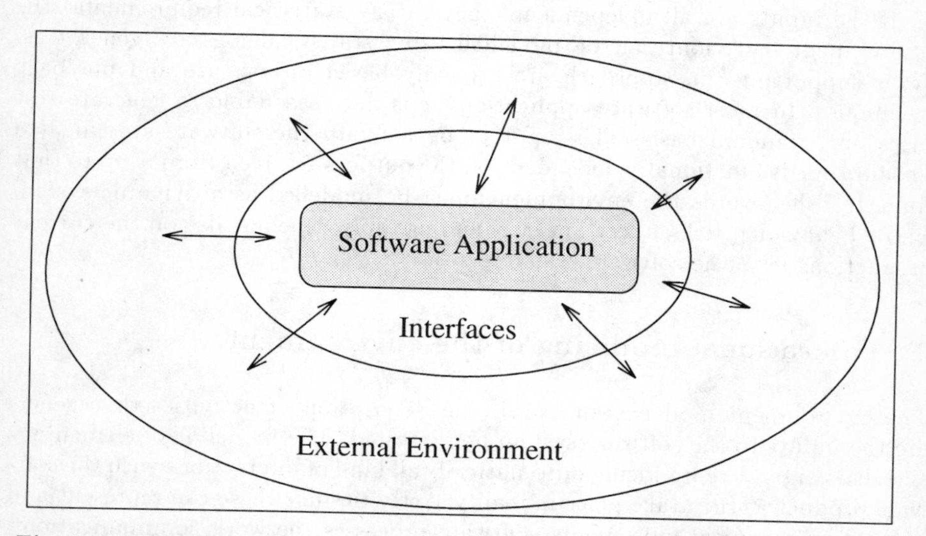

Fig. 3. An emulator models the interactions between environment and application

The environment is a very important concept when dealing with system testing. The environment is defined by considering the closed system comprising the actual software to be tested and all the objects that influence directly or indirecly its inputs.

Validating a software system always implies the definition of an environment which is an abstraction of the real one. We call this abstraction a model of the actual environment: any kind of software validation tool or technique implements a certain model of the actual environment.

The technique of building artificial environments is particularly effective when software systems have complex interactions with their environment. A complex environment is a dynamic system driven by discrete events originated both by itself and by the interaction with the software product and the users. Such environments are typical in supervision systems, distributed applications, client-server architectures, monitoring systems.

The complexity of an environment is related to the rules necessary for defining its dynamic evolution as well as the inputs to the software system being tested. The number of rules and the level of coupling among them indicates the complexity of the environment.

Modeling the environment involves defining communication means (with the product to be tested), the information to be exchanged (messages) and timing constraints (establishing when an interaction occurs).

The interaction between the product and the environment always implies some kind of information exchange in a direct or indirect way. A simple case of indirect interaction is the access to a file. The contents of the file provide the input to the software being tested; in this case the interaction is simple as the reading is completely dependent on the tested software's initiative. Other means are: interprocess communication, networking, driver emulation.

If the inputs are all independent (they satisfy statistical requirements) the environment behaviour can be modelled using statistical message generators. Tools supporting this approach are typically based on capture and playback techniques. In most software applications it is not reasonable to generate test cases on a random basis. The input to be fed into the software system at a certain time is functionally dependent on the outputs received from it up to that time. In other words the environment must be modelled as a dynamic system whose behaviour sticks to certain laws and possibly depends also on the current interactions with the software system.

3 Operational modeling of the environment

The environment model is an executable description of activities that generate the inputs to the software system being tested. The modeling of elementary activities aims at reproducing automatically all kind of interactions with the software product as they take place in reality. Activities like these can represent and simulate mechanical movements, software processes, network communications, user actions. The model must contain the description of all the components that are necessary to express adequately the behavior of the environment system. Therefore elementary activities must be appropriately linked together to express the dynamic interactions taking place within the environment system as well as the related flow of messages.

All activities must be not only described but also executed. Using this approach the requirements for supporting the definition and implementation of the test cases are represented in an operational form.

The executable model allows functional coverage testing as well as performance testing to be performed in highly realistic workload conditions which are determined by generating a number of messages and events (on the basis not only of timing constraints but also of a variety of data values). This should be regarded as the main property of the proposed technique.

We call *emulator* an executable representation of the application's environment, which is used for system testing purposes.

A fundamental role is played by the particular modeling approach used which in this case is based on the Protob [2] [3] language and its support toolset, Artifex.

The graphical Protob formalism proved to be very accessible. The structure of both software and emulator is based on objects mirroring the physical structure of the plant. Since the representation is graphical, understanding is much easier than with an object aggregation hierarchy that is represented textually. Therefore thanks to graphical navigation the tester can easily reach the object icon of an instance of a certain kind of equipment in order to observe its state as well as call its services.

Protob is described in mode detail in another paper [4] included in these proceedings.

4 Assessment of actual applications

It is highly illustrative to analyse the experience of users in applying the validation techniques described here to the development of important projects. We selected three different cases in the area of manufacturing support applications. They concern the issues of supervision and control, planning and monitoring, data-base access and management. In addition two of them refer to system testing and regression testing and one also deals with module and integration testing.

4.1 Fiat Auto: testing of bodyshop supervisors

Fiat Mains is the Information System House of Fiat Auto. The main mission of Fiat Mains is to provide information system support to Fiat Auto including planning, production supervision and logistics.

Such information systems are completely tested in a testing laboratory of Fiat Mains which is specialized in validation service for applications before they are installed in the production plants. Some of the software that is tested in the laboratory is developed internally by Fiat Mains while other applications are carried out by external suppliers. In both cases system testing is performed in the laboratory using the emulation technique.

Figure 4 illustrates a typical architecture of the software systems that are tested in the laboratory.

The experience of the validation laboratory is very wide as more than 20 software systems, such as plant supervisors, production monitors and order management systems, were delivered during the last 5 years of activity. In order to test such applications a number of similar plant emulators were built: they automatically generated all the testing conditions that were expected in the actual environment.

The most valuable benefits reported during this 5-year experience regards the cut down of installation costs and the ability of testing the software even before the actual plant is available or fully operating.

The cost of doing the test with emulators is also a remarkable results. The complete run of the test-cases in the laboratory requires an average of 1 week (regression testing of new releases). Fiat estimated 2 months for an equivalent testing in traditional environments.

The use of emulators brings the highest flexibility in choosing any possible plant condition. Some tests cannot be performed at all on the actual plant whilst they are feasible in the emulated environment (due to costly operations or conditions concerning only operations available in the future). The effectiveness of this technique resulted in completely safe installations. The actual quality of the system is perfectly known after the testing in the laboratory.

A significant case to be reported refers to the supervisor for the new bodyshop plant of Mirafiori, in Torino. The software project was performed by an external supplier and the first releases have been tested and tuned in the laboratory during 3 months in 1993.

The architecture of the emulator is composed of 3 main subsystems: the model of the plant, the interface for the communication with the application to be tested, the interface for the display and control of the plant (figure 4).

The development of the emulator for the Mirafiori plant took 4 months. The effort required for building a new emulator was divided into 3 main activities: implementing the particular material flow and operations by composing building blocks, designing and developing the graphical monitor and control panel for the control of the emulator execution, developing the communication link with the software to be tested.

The emulator model provided the same viewpoint as the actual end-user in charge of controlling the production on the actual plant. The tests have been driven exactly as they were performed in the actual environment. It was very simple to introduce perturbations to the normal flow of the tests thus generating a wide variety of realistic operating conditions. The result of testing with the emulator gave an effective help for validating both the software application and the material flow logic.

During the testing with emulators more than 140 different anomalies were found in the execution of the application and it was possible to fix them before

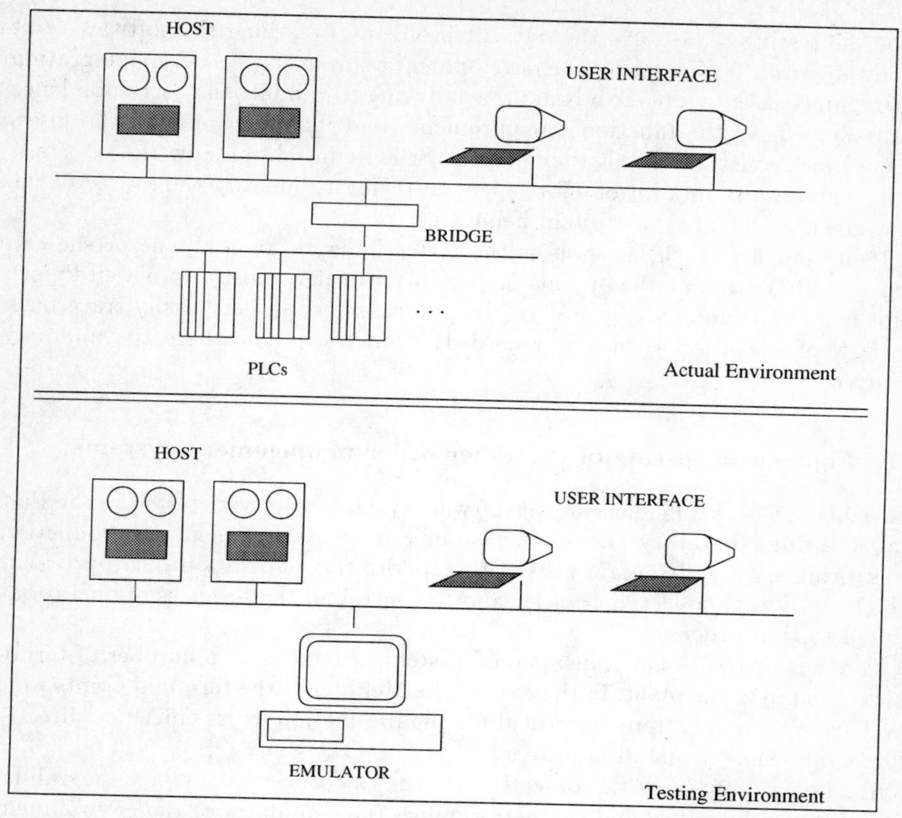

Fig. 4. The architecture of the application and of the testing environment

the installation in the real environment. No more problems were detected in the application once it was operating in the actual plant.

It was realised that stand-alone plant simulations were able to manage only a limited variety of possible threads. The combined execution of the plant model and the controlling software gave the highest possible confidence in the quality of the whole system.

Tests were both controlled interactively and run automatically. Automatic runs facilitated the validation of the largest part of the functionalities whilst the controlled execution of the emulator allowed the software to be debugged during the test.

The emulator of the Mirafiori plant was also used to train the plant operators to the new software system. In particular the new supervisor implied the introduction of a new production management logic characterised by a tight pull policy. The emulator provided a suitable and safe training environment for the people responsible of the planning and scheduling.

A great advantage of using simulation in integration testing is the way it helps the testing team solve the difficult problem of tracking the software fault given the error. The people of the development team in charge of the integration testing must stick to a black box approach, trying to validate the correspondence of the system to the functional requirements and its compliance to the given performance constraints, but they have no or little insight on how the system is built. This results in a lot of effort spent in trying to identify the source of the error given an unexpected system behaviour.

Using our approach, as soon as the tester finds an anomaly, he or she can stop the simulation at that point, or possibly run it again up to a well-known point before the anomaly occurs, so the programmer can very easily trace back the flow of execution (which is recorded) from the point where the anomaly occurred.

4.2 Fiat Auto: testing of customer order management systems

Giove belongs to a different class of software systems that were tested in the Fiat Mains testing laboratory. Giove is a customer order management and production monitoring system. Its main goal is to support the association of a particular vehicle body with a customer order and to control all the progress of operations related to that order.

Giove is a traditional transactional system connected to a number of terminals spread over the plant. In the case of the Melfi plant the terminal points were 26. This system is extremely critical for the production as its efficiency directly affects the whole production process.

In the case of Giove the objective of the test concerned mainly the validation of the architecture and the performance. The emulator of the environment addressed the automatic generation of transactions. The 95% of the transaction workload of the test could be produced automatically by the emulator, reproducing the manual activity of the operators at the terminals. The advantage of using the emulator was that the workload was perfectly consistent with a realistic production scenario. It was possible to test high production rate conditions as well as normal ones.

The cost of developing the emulator in this project was extremely limited, in fact it was equivalent to the cost required for a week of manual test of the same application.

4.3 Cap Volmac: testing a Material Handling Information System

Cap Volmac, the Dutch branch of Cap Gemini Sogeti, has developed the MAHIS control software for a large highly automated factory for coated polypropylene film in the Netherlands. MAHIS, Material Handling Information System, is a mission critical, plant control system, controlling a large automated production plant and a storage warehouse, for coated polypropylene film.

The production process is highly automated and fully under control of MAHIS. Production is ongoing for 24 hours a day, for over 360 days in a year. The MAHIS system is a level-two control system in a three level control hierarchy. It receives production and distribution orders from the logistic and planning level-three control software located on the company's mainframe and commands and co-ordinates the level-one control systems of all individual production machines, the automated guided vehicles (AGV's), the conveyor belt systems and various automated warehouse cranes, which transport the products in various phases of completion through the plant. MAHIS was developed with the Artifex and Quid tools using the Protob formalism.

The positive result experienced by CAP Volmac is the following: developers can produce an operational (executable) model of the module they are developing very early and keep it operational while they are enhancing and refining its functionality.

This approach makes it possible to demonstrate partly functioning prototypes of system parts to the customer very early in the development life cycle, which proves to be very reassuring and a big help in the removal of misconceptions about the intended functionality and behavior. It also has a great influence on module testing: in fact, instead of performing module tests at the end of a development period, developers carry out module testing incrementally at the end of each refinement cycle.

The in-house testing of MAHIS was performed by an independent testing team. They used primarily black-box testing, checking the functionality of the system against its functional specification. The testers were people with expertise in testing and specifying systems at a functional level. They had not taken part in the development nor were trained in the Protob formalism. Nevertheless they could easily comprehend the graphical Protob diagrams of the simulation environment used for testing so well that soon they were able to present problem reports, which clearly pointed out not only the functional defects but also where the cause was located in the system and even contained suggestions for their correction. The formalism proved to be very accessible to them.

The production plant was constructed concurrently with the development of the MAHIS control system. In order to test the control software, a plant simulator was developed to be used as a plant emulator for the tests and it was connected to MAHIS instead of the actual plant under development. Furthermore this approach allowed the control software to be very easily coupled with the plant simulator.

Because MAHIS is a real-time event driven system which has to control the equipments in the plant, it contains a great deal of error handling functions. Typically they include all kinds of time-out management activities. In order to test timing constraints the technique of real-time simulation was used. In this case when the model is executed the delays associated with it give rise to actual delays carried out with the help of the underlying operating system.

The object-based structure of the PROTOB model allows building-blocks representing complete and autonomous pieces of functionality to be defined in conformity with the structure of the equipments in the environment that MAHIS controls. For this reason it was feasible to incrementally deliver MAHIS, in a number of planned releases, instead of performing a single "big-bang" delivery of the whole system. This had a very positive effect on the customer's acceptance testing. The customer was able to spread the acceptance tests over a longer period of time, each time focusing on a specific equipment, with the aid of the supplier of that kind of equipment, and its level-one control software. This made the acceptance testing easier to manage while reducing the peak load on the acceptance test team, which could be made smaller.

In the maintenance phase other people less experienced than the original developers got the task of correcting the remaining problems, further enhancing the functionality and performing regression testing in a simulation environment before implementing the changes in the production environment. The plant simulator, developed for the in-house testing was used for this purpose. They had only a very limited possibility to ask the developers for advice and assistance. In all but the most tricky cases, they were able to cope with this very well. Typically they were able to isolate the cause of a problem or a place into which a functionality enhancement might be introduced very quickly. As a result maintenance costs have been very low so far.

The test configuration, in which a copy of MAHIS is connected to the plant simulator, was also used for the purpose of operator training.

Both the MAHIS development and testing experiences with the PROTOB formalism were very positive. The MAHIS system has been operational for over a year.

Only a few minor problems were encountered during that period. The positive experiences with maintenance so far warrant the expectations, that the choice of the this approach will also have a positive effect on the cost of maintenance in years to come.

5 Conclusions

The results provided by the three projects described in this paper are very clear in expressing the effectiveness of the testing approach based on emulators. In all cases the effort required to develop the emulator was widely paid back by the savings in the execution costs. In addition the use of emulators allowed Fiat Auto to run laboratory test cases that otherwise could have been tried only at the plant site with extremely high risks and costs. Emulators allowed the testing of realistic workloads based not only on the frequency of events but consistent with the actual behaviour. Cap Volmac also reported that the complete test of the software could be carried out even before the actual plant was available.

The success of these applications can be measured in terms of costs savings. The root of these outcomes is to be found in the emulator approach to the testing problem which starts from the analysis of the environment behaviour. A fundamental role is then played by the particular modeling approach based on the Protob methodology.

References

1. P. Zave. The operational versus the conventional approach to software development. *Commun. ACM*, 27:104–18, February 1984.
2. M. Baldassari, G. Bruno, and A. Castella. PROTOB: an object-oriented case tool for modelling and prototyping distributed systems. *Software Practice & Experience*, vol 21(8):823–844, August 1989.
3. G. Bruno. *Model-based software engineering.* Chapman & Hall, London, 1994.
4. G. Bruno and R. Agarwal. Validating software requirements using operational models. In *In these proceedings*, 1995.

Software Testing for Dependability Assessment

Antonia Bertolino

Istituto di Elaborazione della Informazione del CNR
via S. Maria, 46, Pisa, Italy
bertolino@iei.pi.cnr.it

Abstract. Software testing can be aimed at two different goals: removing faults and evaluating dependability. Testing methods described in textbooks having the word "testing" in their title or more commonly used in the industry are mostly intended to accomplish the first goal: revealing failures, so that the faults that caused them can be located and removed. However, the final goal of a software validation process should be to achieve an objective measure of the confidence that can be put on the software being developed. For this purpose, conventional reliability theory has been applied to software engineering and nowadays several reliability growth models can be used to accurately predict the future reliability of a program based on the failures observed during testing. Paradoxically, the most difficult situation is that of a software product that does not fail during testing, as is normally the case for safety-critical applications. In fact, quantification of ultrareliability is impossible at the current state of the art and is the subject of active research. It has been recently suggested that measures of software testability could be used to predict higher dependability than black-box testing alone could do.

1 Introduction

That a certain amount of testing should be included within the software development process is a fact now commonly accepted. However, a general agreement has not been reached on "how much testing" is *enough* and on "which testing method" is *effective*. One reason for this lack of agreement is the fact that one expression, "software testing", is used actually to mean two different things: "correcting" and "measuring". Indeed, testing *enough* or *effectively* is inevitably different depending on which testing goal, correcting or measuring, one is pursuing. In practice, the testing activity does not change, when the two different goals are considered: in both cases, testing consists of checking the behaviour of a program, when it is executed on a finite number of suitably selected test inputs, against the specified expected behaviour. However, the two different goals imply two different (both in size and composition) samples of executions as the most appropriate. In the testing aimed at removing faults, one executes repeatedly a program trying to find out as many failures as possible, so that the faults that caused them can be located and fixed. The goal here is dependability improvement via fault removal, and accordingly testing enough will mean that (a reasonable confidence has been reached that) a large enough part of the faults that were in the program has been found. So, the effectiveness of testing methods is measured in terms of their capability to reveal failures. These methods typically involve issues such as appropriately partitioning the input domain into equivalence classes or as determining special, error-prone, input conditions. On the other hand, in the testing aimed at measuring, one executes a program in order to evaluate how dependable (or, conversely, how prone to failure) it is. Hence, testing enough here means collecting enough test results so that accurate statistical inferences about reliability can be drawn. The effectiveness of measuring methods will involve issues such as deriving a realistic approximation of the user operational profile or choosing the most appropriate inference procedure.

According to these arguments, it is quite intuitive that a testing approach that is very good for removing faults can be very bad for dependability assessment and vice versa. Indeed, in a sense, their respective goals make the two kinds of testing antithetical. If one is performing the testing in order to assess the dependability of a developed software product, then the test inputs should be selected trying to reproduce in the test set up those conditions that will be then found in normal operation. On the contrary, if testing is aimed at removing faults, one should focus the testing effort on those points in the input domain that are more likely to be faulty; this happens more easily for points at the boundary of the input domain or for those inputs that handle special and rarely happening operating conditions. These conditions do not necessarily coincide with "normal" conditions, and in fact the most used conditions are often the least fault-prone.

Both kinds of testing should be carried out; it is however important that their specific goals are recognised and testing is planned accordingly in successive steps. A well thought software testing process should thus include the following phases:

1. A phase of testing for removing faults; in this phase, a testing method that assures a high fault revealing power should be followed.
2. As failures are found, the program is modified and the faults that caused the failures are fixed. If fixes are effective, program reliability increases. Since the population of faults determined by the testing method followed until this point decreases, the method becomes less effective as testing proceeds.
3. Testers could then try other test methods that might be more effective in revealing those faults that are still in the program. This can be repeated until they are satisfied that the correcting testing phase may stop (this is often stated in terms of a specified stopping rule being satisfied).
4. Testing is then devoted to assess program dependability. In this phase, if faults are still found and fixed, a growth of reliability will take place. At this point the scenario is different for the two cases of low to moderate reliability or of ultrareliability. In the first case, the evaluation of the product is conducted following one of the several reliability models and testing can stop when a suitable value is reached. On the other hand, for the evaluation of the reliability of very critical, highly reliable software, the normal scenario is that the program to be evaluated is tested in its final configuration and no fault is found. If a fault were found, the program would be fixed and testing would restart from scratch: one cannot be confident that the fix is effective, and, although clearly testing and removing bugs tends to increase reliability, nobody can tell that an individual fix on a certain program will improve its reliability. So, the only

testing session that will lead to the release of the software is one that shows no faults. Eventually, the testers have to be satisfied that the software is reliable enough: testing can be stopped and the software can be released.

In this paper, I shall provide a general overview of the issues concerning the different phases of such a testing process, with emphasis on testing for dependability assessment. For obvious reasons of space, the overview will be sketchy; however, I shall provide appropriate references for deeper reading. In section 2, I shall touch the topic of testing for correction; this is, in my opinion, the easiest part of the testing job, also because testing techniques in this area have been around for almost 3 decades now, and undoubtedly a richer and well-assessed literature is now available. So, in that section I shall point only at which are currently the weakest points. In Section 3, I shall summarise the issues concerned with software reliability. In Section 4, I shall introduce reliability growth models, the more developed area of the field. In Section 5, I shall describe approaches for measuring (not increasing) reliability. In Section 6, I shall discuss why current methods are not effective for the assessment of ultrareliable software. In Section 7, I shall outline recent proposals of using testability measures to obtain higher dependability predictions than black-box testing alone could do.

2 Catching Faults

The testing that is the subject of textbooks or tutorials having the word "testing" in their titles or as is normally practised in the industry is for the most part concerned with *catching faults*: the program is executed a number of times trying to break it and in this attempt to guess which executions could be faulty, different, more or less systematic, methods can be used. There are now several sources, as [2, 3, 21] just to say a few, that provide a good overview of the different techniques one can follow when testing for revealing failures.

Unfortunately, what sources do not often provide is a valid criterion to choose one method over another. In fact, only recently the evaluation of the effectiveness of testing techniques in detecting faults has began to be addressed in a rigorous, scientific way, e.g., [7, 8]. In the early testing literature, the superiority of one method over another has been often advocated on the basis of unjustified claims or of anecdotal evidence. For example, it was believed by many (e.g., see [21], pg. 36) that systematic testing methods, such as functional methods or coverage-based methods, were much superior to detect failures than random testing, i.e., exercising the software over a purely randomly chosen set of test inputs. However, some rigorous studies [6, 11] have now shown that the two approaches can be equally effective. These studies compared the respective probabilities of failure of a given program when the test inputs were selected randomly over the whole input domain or when instead each test input was taken from within a different class of the domain, after having partitioned it into a number of classes. Indeed, every systematic test method, be it functional or structural, can be thought of as: *i*) partitioning the input domain into classes, such that the program is assumed to behave equivalently on all the inputs of a class w.r.t. the criterion followed, and then *ii*) taking a representative input, or set of inputs, from within each class. For instance, in branch testing, partitions are derived so that every input within a class will exercise one given branch. A deeper analysis provoked by such apparently counter-intuitive results, "no selection criterion can be as effective as any other well-thought one", brought further insight: what possibly makes a test method more effective than another in catching faults is its superior capability to isolate small input classes with a high fault density, when partitioning the input domain. When the partitions induced by a test selection method are not of this kind, but the failure-causing inputs will result almost homogeneously distributed between the derived classes, then taking an input ad hoc from within a class or instead picking it randomly over the whole input domain does not make a substantial difference.

In conclusion, if one is testing a program because he wants to take away faults, then an effective attitude would be that of "suspicion" testing [11]: trying to guess which parts of the input domain are more likely fault-prone, because recently modified, or because the most complex, or for whatever reasonable *suspect*, and concentrating the testing effort on those input parts. However, it must be understood that such approaches, though they can be effective in finding failures, do not provide any meaningful information about the dependability of the tested program. In fact, the *remaining* faults do determine the future failure behaviour of the program, regardless of how many have been already found and removed; and testing for finding failures can never reach a definitive conclusion that the fault found was the last one that was there. So, if testing is aimed at assessing program dependability, the approach of suspicion testing, trying systematically to guess the failure-causing inputs, is not useful. Which approaches can be followed to assess program dependability by testing is the subject of the rest of the paper.

3 The Concept of Software Reliability

A piece of software is dependable if it can be depended upon with justified confidence [15]. Quantification of this concept is very difficult. However, one interesting measure of dependability for a software user is how frequently that software is going to fail. In fact, *reliability* is one of the several factors, or "ilities", into which product quality has been decomposed to be able to measure it, and by far it is the factor for which today the most developed theory exists. Strictly speaking, software reliability is the probability that the software will execute without failure in a given environment for a given period of time [20]. However, this is one of a number of indicators that can be used to measure the failure-proneness of a software product, or its reliability in a wider sense. Other commonly used reliability indicators include: i) MTTF, the expected time interval between current time and the next failure occurrence; ii) MTBF, the expected time interval between one failure occurrence and the next one (this will possibly include the necessary repair time); iii) the failure rate (or ROCOF), the rate at which failures are expected to occur. These measures are probabilistic in nature. In fact, software testing for dependability assessment is an experimental exercise. A single test execution is seen as an experiment in which an input [1] is drawn randomly from a given distribution. In particular, to assess the reliability "in a given environment", this input distribution should approximate as closely as possible the operational input distribution for that environment.

The program execution model underlying this approach is that the input domain contains some failure-causing inputs and the reliability of the software depends on how much frequently those failure-causing inputs will be exercised[2]. The uncertainty in this testing experiment arises from the inputs that will be given to the tested program: in other words, the reason why we talk about the reliability of a software in probabilistic terms is that we can never know which inputs, and in which order, will be executed in operation.

If we introduce the random variable T representing the (uncertain) time to next failure, then the reliability function (conditioned on a specified input distribution) is given by:

$$R(t) = P(T > t), \tag{1}$$

i.e., $R(t)$ is the probability that the software will not fail up to time t. Alternatively, the random variable T can be represented by its distribution function:

$$F(t) = P(T \leq t) = 1 - R(t), \tag{2}$$

commonly called "failure probability". Sometimes, as a synthetic measure of reliability the median of the random variable T is used. It is defined as the value $\bar{t}$ such that:

$$F(\bar{t}) = R(\bar{t}) = \frac{1}{2} \tag{3}$$

A measure of the reliability indicator of interest is obtained by analysing the test results with regard to a selected reliability model. There are two distinct classes of reliability models that one can use: these are referred to as *reliability growth* models and *reliability*, or statistical, models. These two classes of models cover respectively the two situations of programs that are being repaired as failures are encountered and of programs that are not. The latter situation is to be meant

[1] The life of a program is a series of invocations. During each invocation, the program interacts with the rest of the world, receiving inputs and, as a result, producing outputs. We consider all the information received by a program during an invocation as one item, called the input for that invocation. Likewise, we define an output. More precisely, for a "one-shot" program, which reads a vector of values once and runs until it terminates producing a vector of result values, input and output will designate these two vectors. For a program with memory, where "an invocation" may cover many iterations of reading data, modifying the program's internal state and producing outputs, one can define an input to be the whole sequence of data read and an output the whole sequence of data produced, possibly including their timing. Alternatively, one can define as an input both the data read from outside the program and the initial value of its internal state (if the latter is defined in the specification of the program).

[2] It goes without saying that the number of these failure-causing inputs has not a direct correspondence with reliability: a software with many failure-causing inputs that will rarely be exercised is in fact more reliable than a software with just a few ones, but often executed.

as "deferred repair" [20], i.e., the fixes will be incorporated in the next release of the product under test; or, the program has matured enough that fixes are not expected to be beneficial. Models in this class at first glance seem to constitute the simpler case; however, there are some fundamental problems in the underlying assumptions and in their usability. These will be discussed in Section 5. Models in the first class, "with repair", are used to model the situation that a program is tested until a failure is encountered, the fault originating this failure is found and fixed, and the program is again submitted to testing; this process continues until it is evaluated that an adequate reliability has been reached. These models are called reliability growth models, understanding that as the software undergoes testing and is repaired, its reliability increases. Testing with repair and reliability growth models will be discussed in the next section.

4 Reliability Growth Models[3]

The first software reliability growth model [14] appeared in the early 1970's. Today, tens of models are in use. There are fundamentally two categories of them:

1. Times Between Failures Models, in which the random variable of interest is the time between failures.
2. Failure Count Models, that are concerned with the number of failures detected in specified time intervals.

However, the two categories are strictly related and to some extent it is possible to switch from each other. In the following, we provide a description of the basic concepts underlying inter-failure times models.

Given a program p, when, by testing it, we detect a failure, we attempt to remove the fault that caused it by modifying p: this has the effect of obtaining a new program p_1. Continuing this testing and repair process, we shall obtain a sequence of programs $p_1, p_2, \ldots, p_n$. Let $T_1, T_2, \ldots, T_n$ be the random variables associated with the successive inter-failure times, i.e., T_i denotes the time between the $(i-1)$st and ith failure. At the current time, the tester will have available the sequence of *observed* inter-failure times, $t_1, t_2, \ldots, t_{(i-1)}$: these observed times are realisations of the random variables $T_1, T_2, \ldots, T_{(i-1)}$. The assessment of the current software reliability is actually a *prediction*: from the observed inter-failure times, we want to infer the *future* realisation of unobserved T_i. Hence, more precisely, reliability growth models are *prediction systems*: they provide the tester with a probabilistic model of the distributions of the T_i's under some unknown parameters and a statistical inference procedure to infer adequate values for the unknown parameters using the observed data. Then, plugging the inferred parameter values into the model, a prediction about T_i can be derived.

We already noted in the previous section that the source of uncertainty in the probabilistic model of software failure behaviour is the sequence of executed inputs. Some reliability growth models, e.g. [19], introduce as a second source of

[3] The contents of this section are almost entirely derived from [17], to which we refer for deeper reading.

uncertainty the effectiveness of fixes. In fact, the assumption that after-failure repairs are always successful might be unrealistic: on the contrary, sometimes a program might be made less reliable as a result of an attempt to remove a fault.

Researchers agree that no "universally best" model can be individuated; a recent study [1] has shown that although a few models can be rejected because they perform badly in all situations, no single model always performs well for all potential users. However, at the current state of the art, it is usually possible to obtain reasonably accurate reliability predictions. What the users of reliability growth models should do is to try as many as possible models on their data source, and determine which is the most accurate for their need. From a set of models, the tester can derive a reliability prediction based upon $t_1, t_2, \ldots, t_{(j-1)}$ and then compare the various predictions with the already observed t_j, for a number of collected failure times $t_j, j < i$. Typically, he will observe a great disagreement between the different predictions. He will then choose the model that has given the most trustworthy predictions for the past data.

There are two problems that must be considered in this process of choosing the best model for the data at hand. The first involves bias, and arises when a prediction system provides inferences that are consistently wrong, either towards the optimistic side (the predicted reliability is always higher than the effective reliability) or towards the pessimistic side (the predicted reliability is always lower than the effective reliability). The other problem is noise of the prediction: how the predictions fluctuate around the real values.

Besides, even choosing that model that gives the best predictions on the observed data after such an accurate analysis is not a priori guaranteed to work: this approach is based on the assumption that there is a continuity in the failure manifestation between past and future. This consideration, also, confirms the importance of ensuring that testing for reliability assessment take place in an environment that accurately reflects the conditions under which the software will actually operate. Indeed, the main obstacle to the use of these models in software development remain the problem of reproducing as closely as possible the conditions of operational use. This problem is common to all reliability models and will be further discussed in the next section, in which we speak about reliability testing without repair.

5 Life Testing

Reliability, or statistical, models are used after the debugging process is finished to assess the reliability one can expect from the software under test. The basic approach is to generate randomly a set n of test cases from an input distribution shaped to approximate closely the *operational* distribution. A reliability measure is then derived based on the number of failures observed in the execution of the n sampled inputs. If f is the number of inputs that have raised failures, then an estimate of reliability [22] is very simply given by

$$R = 1 - \frac{f}{n} \qquad (4)$$

However, testing for reliability assessment is generally conducted when a program does not fail anymore, and so there are not observed failures against which the predicted reliability can be checked. Using a purely statistical estimate, we could evaluate the probability that the reliability is high enough, or, conversely, that the failure probability is below a specified upper bound. For this purpose, the following assumptions are generally made: during the testing process, n inputs are drawn randomly and with replacement from a specified input distribution; the probability of failure per input ϑ is constant over time; each test outcome is independent from the others. Under these assumptions, the probability that a single test will fail is ϑ, or, conversely, $(1 - \vartheta)$ that it will not fail. Therefore, the probability that the software will pass n tests and yet its failure probability is higher than ϑ is not higher than C, with:

$$C \leq (1 - \vartheta)^n \tag{5}$$

This is often stated as meaning that "there is a confidence (1-C) in the estimate of ϑ as an upper bound for the *true* failure probability of the software".

This conclusion derives from the application to software of conventional reliability theory, that is based on a "frequentist" definition of probability. Some researchers question the plausibility of such interpretation of probability for the assessment of dependability and suggest instead that using a Bayesian, subjective, interpretation is more satisfactory [16]. In the Bayesian approach, probabilities are not related with the frequency of occurrence of certain events, but describe the strength of an observer's belief about whether such events will take place. It is assumed that the observer has some prior belief that will change as a result of seeing the outcome of the "experiment", in which the data are collected. Bayesian theory provides a formalism for updating, using experimental evidence, the"prior belief" held before observing this evidence. Using a Bayesian approach, the (posterior) reliability function after having observed no failure is (details of the derivation procedure can be found in [18]):

$$R(t|no\, failures\, up\, to\, time\, t_0) = \left(\frac{b + t_0}{b + t_0 + t}\right)^a \tag{6}$$

where a and b are parameters representing the observer's prior belief about the failure rate. In particular, for an observer with a certain form of complete "prior ignorance" [18]:

$$R(t|no\, failures\, up\, to\, time\, t_0) = \left(\frac{t_0}{t_0 + t}\right) \tag{7}$$

In [9], Hamlet argues that the software analogy to mechanical reliability is a poor one and that relying on software reliability figures predicted according to reliability models (either conventional or Bayesian) can be dangerous. He explains that for software many of the assumptions underlying the current reliability theory are flawed. The most significant deficiencies are: we cannot assume an operational profile for software, and test executions are not always independent. Therefore, testing by sampling the input domain is not meaningful: apparently independent inputs can lead to a same program state and possibly to a same fault, and thus they are not independent samples. Indeed, he points out the necessity to develop a "new", more plausible, dependability theory [10]. The sampling space for this new theory could be the program state space. Besides, rather than trying to estimate the reliability of the software, the new theory would try to estimate the probability that a piece of software is perfect. Attempting to predict absence of faults is of course aiming higher than attempting just to predict an acceptable failure rate. However, its advantages are:

1. (perhaps obviously) knowledge can be brought to bear that is not otherwise used when predicting failure rates, i.e., the knowledge that without faults no failure is possible;
2. the testing profile does not need to reproduce an operational distribution and can be chosen so as to best detect faults (i.e., so as to improve failure detection effectiveness);
3. by not conditioning the prediction on an input profile, one could state a reliability figure for the software when used in any condition within the range for which it is specified, just as done for off-the-shelf hardware;
4. when requiring high reliability over long periods of continuous operation, reasoning in terms of probability that the product is fault-free may yield more favourable predictions than reasoning from probabilities of failure per execution.

6 The Problem with Safety-critical Software

Software systems are increasingly used in life-critical applications, for which reliability requirements are extremely high. In the region commonly called of *ultrareliability*, systems must be designed to obtain a probability of failure on the order of 10^{-7} to 10^{-9} per hour. Such reliability levels are several orders of magnitude beyond those that can be validated at the current state of the art, using one of the models described till now. Justifiably claiming before operation (as required by safety regulations), on the basis of pure black-box testing, that such reliability requirements have been achieved is impossible in practice.

For example, setting $t = t_0$ and substituting in 7,we have:

$$R(t_0) = \tfrac{1}{2},$$

i.e., having observed a period t_0 of failure-free working, we have a 50% probability of observing a further period t_0 without failure. In other words, if we need to estimate a posterior median[4] time to failure of, say, 10^9 hours, and we do not bring any prior belief to the problem, we would need to see 10^9 hours of failure-free working. So, if we try to reach such an ultrareliability assessment using only information obtained from the failure behaviour of the software tested as a black-box, then we would need to test the system for a very long, prohibitive, time.

[4] Remember that the median is defined as that value $\bar{t}$ such that $F(\bar{t}) = R(\bar{t}) = \frac{1}{2}$

In [5], evidence is given that quantification of life-critical software reliability using statistical methods is infeasible whether applied to standard software or fault-tolerant software. So, the conclusions of both [5] and [18] are that validation of ultrareliable software cannot be done based solely on black-box testing, but other forms of quantifiable evidence must also be collected.

7 Use of Testability Measures

A recently suggested approach towards raising the level of reliability that can be assessed by testing is to reason directly about the effectiveness of testing, assuming that we can measure the *testability* of the program considered. The term "testability" has been long used informally to capture the intuitive notion of how easily a program can be tested. However, here, by testability we mean a precise, mathematical concept, introduced by Voas [23] as the conditional probability that a program will fail under test, if it contains faults. When dealing with testing, we must take into account that test outcomes are observed by a mechanism, called *oracle*. The effectiveness of testing thus depends heavily on the used oracle. On one hand, the oracle itself could not be perfect, but could produce an incorrect judgement. This is a reasonable objection when many test outputs have to be automatically checked. On the other hand, an oracle could be designed to analyse not only failures at program output, but also errors in the program's internal state, for example by instrumenting the program with executable assertions for self-checking.

So, in [4], the following, more rigorous definition of testability was given:

the testability of a program, *Testab*, is the probability that a test of the program on an input drawn from a specified probability distribution is rejected, given a specified oracle and given that the program is faulty.

The notion of testability can help to guide the testing process [24]. For example, it can be sensible to concentrate more testing on modules with lower testability, or to estimate how many tests are necessary to achieve a desired confidence in software correctness [13].

Recently, Hamlet and Voas [12] have suggested to use testability-based models as a way to "amplify" the results of testing, i.e., to predict higher reliability than that that can be directly measured by testing alone. More specifically, they state that the test results can be interpreted in the light of additional knowledge about the program under test acquired by "sensitivity analysis". Precisely, they give a method for estimating the probability that a program contains no residual faults.

Agreeing with their intuition, but finding some problems with the presentation in [12], in [4] we have developed another, more rigorous inference procedure for the probability of correctness, using a Bayesian approach. Suppose that we have a measure of the testability of our program, *Testab*, and besides that our prior belief in the perfection of our program is P_p. After observing no failures or errors in n tests, our posterior probability (i.e., our updated belief) of the software being correct is [4]:

$$P(nofault|n\, successful\, tests) = \frac{P_p}{P_p + (1 - P_p)(1 - Testab)^n} \qquad (8)$$

This formula shows the importance of both the estimated testability and the prior belief in the software being fault-free in shaping the posterior belief. The prior belief itself could be derived, in our case, by considering experience with previous, comparable, software products, and the static verification procedures applied on the product to be evaluated (e.g., formal proofs, inspections), in the light of previous experience with their effectiveness.

We also discuss the claim of [12] that a high testability is a desirable property for a program. This claim seems counter-intuitive: most programs contain bugs, and they are the more useful the less often these bugs produce failures. It is true, of course, that after a series of successful tests a high testability implies a high probability that no faults remain in the tested program. But it also implies a high probability of failure for the remaining faults, if any.

In fact, an increase of the testability has two contrasting effects on the reliability of a tested program: it increases the confidence in the absence of faults, but, on the other hand, it reduces the robustness of the program, if faults do remain. Using 8, in [4] we derived an expression of the probability of failure ϑ vs. $Testab$. We could then study the effect of a change in $Testab$ on ϑ. We obtained a family of curves, that all have only one maximum for a certain value of $Testab$, depending on the number of successful tests and on the prior probability of perfection. Thus, as $Testab$ increases, ϑ decreases only if the starting value of $Testab$ was higher than the value by which ϑ is maximum. If $Testab$ was lower than this value, then an increase of $Testab$ actually provokes an increase also of ϑ, i.e., for low values of testability the negative effect of increasing it will offset the positive one.

8 Conclusions

A software validation process is meaningful only if it can produce an objective assessment of the dependability of the software product under development. One interesting measure of dependability for the software user is how frequently that piece of software is going to fail, or its reliability. For moderate reliability requirements, accurate reliability predictions can be achieved during debug testing using one of the several reliability growth models today available. Obviously, as the level of required reliability increases, its assessment requires progressively more effort. So, quantification of ultrareliability, as would be necessary for life-critical application, is currently infeasible, not because existing methods wouldn't be applicable in theory, but because in practice prohibitively long testing sessions would be needed.

On the other hand, application of conventional reliability theory to software engineering has been often criticised as being not meaningful. The toughest problems derive from the underlying assumptions that test inputs are drawn from the operational distribution and that different test outputs are independent.

A "new" dependability theory for software is highly needed. Considering the problems cited above, it could be more useful to assess the probability that a program is correct, instead of its probability of failure. Software testability could play a role in this new theory. Testability is certainly a useful notion in describing the factors of interest in the testing activity. A high testability can be desirable, because it increases the likelihood of absence of faults, after a series of successful tests, but it can also characterise very "brittle" software.

The idea of "reliability amplification" can help to solve the problem of ultrareliability quantification, because it puts together the observation of software failures under reliability testing with the analysis of software defects under testability evaluation. A recently proposed Bayesian approach has been outlined, that uses measures of testability to derive both the probability of absence of faults and the probability of failure. However, these results are still preliminary and certainly need further study: the main points to be explored are the fault/failure model and how testability can be evaluated.

Acknowledgements

This paper includes some recent work conducted with Lorenzo Strigini within project "SHIP", funded by the European Commission in the framework of the Environment Programme.

References

1. Abdel-Ghaly, A. A., Chan, P. Y., Littlewood, B.: Evaluation of Competing Software Reliability Predictions. IEEE Trans. Software Eng. **SE-12**(9) (1986) 950-967
2. Adrion, W. R., Branstad, M. A., Cherniavsky, J. C.: Validation, Verification and Testing of Computer Software. ACM Computing Surveys. **14**(2) (1982)
3. Beizer, B.: Software Testing Techniques, Second Edition. Van Nostrand Reinhold, New York. 1990
4. Bertolino, A., Strigini, L.: Using Testability Measures for Dependability Assessment. Proc. 17th Intern. Conference on Software Engineering. Seattle, Wa. April (1995).
5. Butler, R. W., Finelli, G. B.: The Infeasibility of Experimental Quantification of Life-Critical Software Reliability. IEEE Trans. Software Eng. **19**(1) (1993) 3-12
6. Duran, J. W., Ntafos, S. C.: An Evaluation of Random Testing. IEEE Trans. Software Eng. **SE-10**(4) (1984) 438-444
7. Frankl, P. G.,Weyuker, E. J.: A Formal Analysis of the Fault Detection Ability of Testing Methods. IEEE Trans. Software Eng. **19**(3) (1993) 202-213
8. Frankl, P. G., Weiss, S. N.: An Experimental Comparison of the Effectiveness of Branch Testing and Data Flow Testing. IEEE Trans. Software Eng. **19**(8) (1993) 774-787
9. Hamlet, D.: Are We Testing for True Reliability? IEEE Software. July (1992) 21-27
10. Hamlet, D.: Foundations of Software Testing: Dependability Theory. Proc. 2nd ACM SIGSOFT Symp. Foundations Software Eng. New Orleans, USA. December 1994. In ACM SIGSOFT **19**(5) (1994) 128-139
11. Hamlet, D., Taylor, R.: Partition Testing Does Not Inspire Confidence. IEEE Trans. Software Eng. **16**(12) (1990) 1402-1411
12. Hamlet, D., Voas, J.: Faults on Its Sleeve: Amplifying Software Reliability Testing. Proc. Int. Symposium on Software Testing and Analysis (ISSTA). Cambridge, Massachusetts. June (1993) 89-98
13. Howden, W. E., Huang, Y.: Analysis of Testing Methods Using Failure Rate and Testability Models. Tech. Report CSE, Univ. of California at San Diego. (1993)

14. Jelinski, Z., Moranda, P. B.: Software Reliability Research. In Freiberger, W. (Ed.): Statistical Computer Performance. Academic Press, New York. (1972) 465-484
15. Laprie, J. C.: Dependability: Basic Concepts and Terminology. Dependable Computing and Fault-Tolerant Systems. Vol. 5. Springer-Verlag, Wien New York. 1992
16. Littlewood, B.: How to Measure Software Reliability and How Not To. IEEE Trans. Reliability. **R-28**(2) (1979) 103-110
17. Littlewood, B.: Modelling Growth in Software Reliability. In Rook, P. (Ed.): Software Reliability Handbook. Elsevier Applied Science, London and New York. 1990
18. Littlewood, B., Strigini, L.: Validation of Ultra-High Dependability for Software-based Systems. Communications of the ACM. **36**(11) (1993) 69-80
19. Littlewood, B., Verrall, J. L.: A Bayesian Reliability Growth Model for Computer Software. J. Royal Statist. Soc., Appl. Statist. **22**(3) (1973) 332-346
20. Musa, J. D., Iannino, A., Okumoto, K.: Software Reliability Measurement, Prediction, Application. McGraw-Hill, New York. 1987
21. Myers, G. J.: The Art of Software Testing. J. Wiley & Sons, New York. 1979.
22. Nelson, E.: Estimating Software Reliability from Test Data. Microelectron. Rel. **17** (1978) 67-74
23. Voas, J. M.: PIE: A Dynamic Failure-Based Technique. IEEE Trans. Software Eng. **18**(8) (1992) 717-727
24. Voas, J., Morell, L., Miller, K.: Predicting Where Faults Can Hide from Testing. IEEE Software. March (1991) 41-48

Author Index

Lecture Notes in Computer Science

For information about Vols. 1–848
please contact your bookseller or Springer-Verlag

Vol. 849: R. W. Hartenstein, M. Z. Servít (Eds.), Field-Programmable Logic. Proceedings, 1994. XI, 434 pages. 1994.

Vol. 850: G. Levi, M. Rodríguez-Artalejo (Eds.), Algebraic and Logic Programming. Proceedings, 1994. VIII, 304 pages. 1994.

Vol. 851: H.-J. Kugler, A. Mullery, N. Niebert (Eds.), Towards a Pan-European Telecommunication Service Infrastructure. Proceedings, 1994. XIII, 582 pages. 1994.

Vol. 852: K. Echtle, D. Hammer, D. Powell (Eds.), Dependable Computing – EDCC-1. Proceedings, 1994. XVII, 618 pages. 1994.

Vol. 853: K. Bolding, L. Snyder (Eds.), Parallel Computer Routing and Communication. Proceedings, 1994. IX, 317 pages. 1994.

Vol. 854: B. Buchberger, J. Volkert (Eds.), Parallel Processing: CONPAR 94 – VAPP VI. Proceedings, 1994. XVI, 893 pages. 1994.

Vol. 855: J. van Leeuwen (Ed.), Algorithms – ESA '94. Proceedings, 1994. X, 510 pages.1994.

Vol. 856: D. Karagiannis (Ed.), Database and Expert Systems Applications. Proceedings, 1994. XVII, 807 pages. 1994.

Vol. 857: G. Tel, P. Vitányi (Eds.), Distributed Algorithms. Proceedings, 1994. X, 370 pages. 1994.

Vol. 858: E. Bertino, S. Urban (Eds.), Object-Oriented Methodologies and Systems. Proceedings, 1994. X, 386 pages. 1994.

Vol. 859: T. F. Melham, J. Camilleri (Eds.), Higher Order Logic Theorem Proving and Its Applications. Proceedings, 1994. IX, 470 pages. 1994.

Vol. 860: W. L. Zagler, G. Busby, R. R. Wagner (Eds.), Computers for Handicapped Persons. Proceedings, 1994. XX, 625 pages. 1994.

Vol: 861: B. Nebel, L. Dreschler-Fischer (Eds.), KI-94: Advances in Artificial Intelligence. Proceedings, 1994. IX, 401 pages. 1994. (Subseries LNAI).

Vol. 862: R. C. Carrasco, J. Oncina (Eds.), Grammatical Inference and Applications. Proceedings, 1994. VIII, 290 pages. 1994. (Subseries LNAI).

Vol. 863: H. Langmaack, W.-P. de Roever, J. Vytopil (Eds.), Formal Techniques in Real-Time and Fault-Tolerant Systems. Proceedings, 1994. XIV, 787 pages. 1994.

Vol. 864: B. Le Charlier (Ed.), Static Analysis. Proceedings, 1994. XII, 465 pages. 1994.

Vol. 865: T. C. Fogarty (Ed.), Evolutionary Computing. Proceedings, 1994. XII, 332 pages. 1994.

Vol. 866: Y. Davidor, H.-P. Schwefel, R. Männer (Eds.), Parallel Problem Solving from Nature - PPSN III. Proceedings, 1994. XV, 642 pages. 1994.

Vol 867: L. Steels, G. Schreiber, W. Van de Velde (Eds.), A Future for Knowledge Acquisition. Proceedings, 1994. XII, 414 pages. 1994. (Subseries LNAI).

Vol. 868: R. Steinmetz (Ed.), Multimedia: Advanced Teleservices and High-Speed Communication Architectures. Proceedings, 1994. IX, 451 pages. 1994.

Vol. 869: Z. W. Raś, Zemankova (Eds.), Methodologies for Intelligent Systems. Proceedings, 1994. X, 613 pages. 1994. (Subseries LNAI).

Vol. 870: J. S. Greenfield, Distributed Programming Paradigms with Cryptography Applications. XI, 182 pages. 1994.

Vol. 871: J. P. Lee, G. G. Grinstein (Eds.), Database Issues for Data Visualization. Proceedings, 1993. XIV, 229 pages. 1994.

Vol. 872: S Arikawa, K. P. Jantke (Eds.), Algorithmic Learning Theory. Proceedings, 1994. XIV, 575 pages. 1994.

Vol. 873: M. Naftalin, T. Denvir, M. Bertran (Eds.), FME '94: Industrial Benefit of Formal Methods. Proceedings, 1994. XI, 723 pages. 1994.

Vol. 874: A. Borning (Ed.), Principles and Practice of Constraint Programming. Proceedings, 1994. IX, 361 pages. 1994.

Vol. 875: D. Gollmann (Ed.), Computer Security – ESORICS 94. Proceedings, 1994. XI, 469 pages. 1994.

Vol. 876: B. Blumenthal, J. Gornostaev, C. Unger (Eds.), Human-Computer Interaction. Proceedings, 1994. IX, 239 pages. 1994.

Vol. 877: L. M. Adleman, M.-D. Huang (Eds.), Algorithmic Number Theory. Proceedings, 1994. IX, 323 pages. 1994.

Vol. 878: T. Ishida; Parallel, Distributed and Multiagent Production Systems. XVII, 166 pages. 1994. (Subseries LNAI).

Vol. 879: J. Dongarra, J. Waśniewski (Eds.), Parallel Scientific Computing. Proceedings, 1994. XI, 566 pages. 1994.

Vol. 880: P. S. Thiagarajan (Ed.), Foundations of Software Technology and Theoretical Computer Science. Proceedings, 1994. XI, 451 pages. 1994.

Vol. 881: P. Loucopoulos (Ed.), Entity-Relationship Approach – ER'94. Proceedings, 1994. XIII, 579 pages. 1994.

Vol. 882: D. Hutchison, A. Danthine, H. Leopold, G. Coulson (Eds.), Multimedia Transport and Teleservices. Proceedings, 1994. XI, 380 pages. 1994.

Vol. 883: L. Fribourg, F. Turini (Eds.), Logic Program Synthesis and Transformation – Meta-Programming in Logic. Proceedings, 1994. IX, 451 pages. 1994.

Vol. 884: J. Nievergelt, T. Roos, H.-J. Schek, P. Widmayer (Eds.), IGIS '94: Geographic Information Systems. Proceedings, 1994. VIII, 292 pages. 19944.

Vol. 885: R. C. Veltkamp, Closed Objects Boundaries from Scattered Points. VIII, 144 pages. 1994.

Vol. 886: M. M. Veloso, Planning and Learning by Analogical Reasoning. XIII, 181 pages. 1994. (Subseries LNAI).

Vol. 887: M. Toussaint (Ed.), Ada in Europe. Proceedings, 1994. XII, 521 pages. 1994.

Vol. 888: S. A. Andersson (Ed.), Analysis of Dynamical and Cognitive Systems. Proceedings, 1993. VII, 260 pages. 1995.

Vol. 889: H. P. Lubich, Towards a CSCW Framework for Scientific Cooperation in Europe. X, 268 pages. 1995.

Vol. 890: M. J. Wooldridge, N. R. Jennings (Eds.), Intelligent Agents. Proceedings, 1994. VIII, 407 pages. 1995. (Subseries LNAI).

Vol. 891: C. Lewerentz, T. Lindner (Eds.), Formal Development of Reactive Systems. XI, 394 pages. 1995.

Vol. 892: K. Pingali, U. Banerjee, D. Gelernter, A. Nicolau, D. Padua (Eds.), Languages and Compilers for Parallel Computing. Proceedings, 1994. XI, 496 pages. 1995.

Vol. 893: G. Gottlob, M. Y. Vardi (Eds.), Database Theory – ICDT '95. Proceedings, 1995. XI, 454 pages. 1995.

Vol. 894: R. Tamassia, I. G. Tollis (Eds.), Graph Drawing. Proceedings, 1994. X, 471 pages. 1995.

Vol. 895: R. L. Ibrahim (Ed.), Software Engineering Education. Proceedings, 1995. XII, 449 pages. 1995.

Vol. 896: R. N. Taylor, J. Coutaz (Eds.), Software Engineering and Human-Computer Interaction. Proceedings, 1994. X, 281 pages. 1995.

Vol. 897: M. Fisher, R. Owens (Eds.), Executable Modal and Temporal Logics. Proceedings, 1993. VII, 180 pages. 1995. (Subseries LNAI).

Vol. 898: P. Steffens (Ed.), Machine Translation and the Lexicon. Proceedings, 1993. X, 251 pages. 1995. (Subseries LNAI).

Vol. 899: W. Banzhaf, F. H. Eeckman (Eds.), Evolution and Biocomputation. VII, 277 pages. 1995.

Vol. 900: E. W. Mayr, C. Puech (Eds.), STACS 95. Proceedings, 1995. XIII, 654 pages. 1995.

Vol. 901: R. Kumar, T. Kropf (Eds.), Theorem Provers in Circuit Design. Proceedings, 1994. VIII, 303 pages. 1995.

Vol. 902: M. Dezani-Ciancaglini, G. Plotkin (Eds.), Typed Lambda Calculi and Applications. Proceedings, 1995. VIII, 443 pages. 1995.

Vol. 903: E. W. Mayr, G. Schmidt, G. Tinhofer (Eds.), Graph-Theoretic Concepts in Computer Science. Proceedings, 1994. IX, 414 pages. 1995.

Vol. 904: P. Vitányi (Ed.), Computational Learning Theory. EuroCOLT'95. Proceedings, 1995. XVII, 415 pages. 1995. (Subseries LNAI).

Vol. 905: N. Ayache (Ed.), Computer Vision, Virtual Reality and Robotics in Medicine. Proceedings, 1995. XIV, 567 pages. 1995.

QMW LIBRARY
(MILE END)

Vol. 906: E. Astesiano, G. Reggio, A. Tarlecki (Eds.), Recent Trends in Data Type Specification. Proceedings, 1995. VIII, 523 pages. 1995.

Vol. 907: T. Ito, A. Yonezawa (Eds.), Theory and Practice of Parallel Programming. Proceedings, 1995. VIII, 485 pages. 1995.

Vol. 908: J. R. Rao Extensions of the UNITY Methodology: Compositionality, Fairness and Probability in Parallelism. XI, 178 pages. 1995.

Vol. 909: H. Comon, J.-P. Jouannaud (Eds.), Term Rewriting. Proceedings, 1993. VIII, 221 pages. 1995.

Vol. 910: A. Podelski (Ed.), Constraint Programming: Basics and Trends. Proceedings, 1995. XI, 315 pages. 1995.

Vol. 911: R. Baeza-Yates, E. Goles, P. V. Poblete (Eds.), LATIN '95: Theoretical Informatics. Proceedings, 1995. IX, 525 pages. 1995.

Vol. 912: N. Lavrac˘, S. Wrobel (Eds.), Machine Learning: ECML – 95. Proceedings, 1995. XI, 370 pages. 1995. (Subseries LNAI).

Vol. 913: W. Schäfer (Ed.), Software Process Technology. Proceedings, 1995. IX, 261 pages. 1995.

Vol. 914: J. Hsiang (Ed.), Rewriting Techniques and Applications. Proceedings, 1995. XII, 473 pages. 1995.

Vol. 915: P. D. Mosses, M. Nielsen, M. I. Schwartzbach (Eds.), TAPSOFT '95: Theory and Practice of Software Development. Proceedings, 1995. XV, 810 pages. 1995.

Vol. 916: N. R. Adam, B. K. Bhargava, Y. Yesha (Eds.), Digital Libraries. Proceedings, 1994. XIII, 321 pages. 1995.

Vol. 917: J. Pieprzyk, R. Safavi-Naini (Eds.), Advances in Cryptology - ASIACRYPT '94. Proceedings, 1994. XII, 431 pages. 1995.

Vol. 918: P. Baumgartner, R. Hähnle, J. Posegga (Eds.), Theorem Proving with Analytic Tableaux and Related Methods. Proceedings, 1995. X, 352 pages. 1995. (Subseries LNAI).

Vol. 919: B. Hertzberger, G. Serazzi (Eds.), High-Performance Computing and Networking. Proceedings, 1995. XXIV, 957 pages. 1995.

Vol. 920: E. Balas, J. Clausen (Eds.), Integer Programming and Combinatorial Optimization. Proceedings, 1995. IX, 436 pages. 1995.

Vol. 921: L. C. Guillou, J.-J. Quisquater (Eds.), Advances in Cryptology – EUROCRYPT '95. Proceedings, 1995. XIV, 417 pages. 1995.

Vol. 923: M. Meyer (Ed.), Constraint Processing. IV, 289 pages. 1995.

Vol. 924: P. Ciancarini, O. Nierstrasz, A. Yonezawa (Eds.), Object-Based Models and Languages for Concurrent Systems. Proceedings, 1994. VII, 193 pages. 1995.

Vol. 925: J. Jeuring, E. Meijer (Eds.), Advanced Functional Programming. Proceedings, 1995. VII, 331 pages. 1995.

Vol. 926: P. Nesi (Ed.), Objective Software Quality. Proceedings, 1995. VIII, 249 pages. 1995.

Vol. 927: J. Dix, L. Moniz Pereira, T. C. Przymusinski (Eds.), Non-Monotonic Extensions of Logic Programming. Proceedings, 1994. IX, 229 pages. 1995. (Subseries LNAI).